Lectures on Fourier Series

Lectures on Fourier Series

L. SOLYMAR

Department of Engineering Science,
University of Oxford

Oxford   New York   Tokyo
OXFORD UNIVERSITY PRESS
1988

Oxford University Press, Walton Street, Oxford OX2 6DP

Oxford New York Toronto
Delhi Bombay Calcutta Madras Karachi
Petaling Jaya Singapore Hong Kong Tokyo
Nairobi Dar es Salaam Cape Town
Melbourne Auckland
and associated companies in
Berlin Ibadan

Oxford is a trade mark of Oxford University Press

Published in the United States
by Oxford University Press, New York

British Library Cataloguing in Publication Data
Solymar, L. (Laszlo)
Lectures on Fourier series.
1. Fourier series
I. Title
515'.2433
ISBN 0-19-856198-9
ISBN 0-19-856199-7 (Pbk.)

Library of Congress Cataloging in Publication Data
Solymar, L. (Laszlo)
Lectures on Fourier series/L. Solymar.
p.   cm.
Includes index.
1. Evolution—Mathematical models. 2. Fourier series. I. Title.
QA404.S65 1988   515'.2433—dc 19   88-17590
ISBN 0-19-856198-9
ISBN 0-19-856199-7 (Pbk.)

Typeset by The Universities Press (Belfast) Ltd
Printed in Great Britain
at the University Printing House, Oxford
by David Stanford
Printer to the University

Preface

THIS book is an expanded version of my lectures given in Hilary term 1987 to first year undergraduates in the Department of Engineering Science, University of Oxford. I aimed at a text that is readable and offers an easy introduction to the subject with the aid of a large number of examples. I hope it will be of interest not only to undergraduates but to applied scientists as well who could look up formulae, refresh memories, and might even come across techniques (*e.g.* how to solve a non-linear differential equation with periodic excitation) useful for their present work.

I am greatly indebted to Dr. R. R. A. Syms of Imperial College, London and Dr. C. J. Budd of Hertford College, Oxford for reading the first version of the manuscript and for useful comments.

Oxford   L.S.
June 1988

Contents

Introduction

A CHARACTERISTIC of applied sciences is that very often the relationships between variables can be put in mathematical form, and this is particularly true for engineering. This means, as you may have realized by now, that you need to study mathematics, a fair amount of mathematics. Some of it is quite interesting, some other parts of the subject may have only limited appeal. I believe Fourier series belongs to the part of mathematics that is not only interesting, not only full of excitement and surprises, but happens to be useful as well. Nature is full of periodic phenomena, and Fourier series are absolutely indispensable in describing periodic phenomena to various degrees of approximation.

I contend—well, I hope—that by unravelling the secrets of this branch of mathematics the following lectures will radically change your views about approximations in general and about approximating periodic functions in particular. You will learn new concepts and a powerful technique. I would actually go further and say that it is one of the most powerful techniques ever invented for the benefit of applied scientists.

People will dispute whether the man who gave his name to the series, Jean Baptiste Joseph Fourier (not to be confused with his contemporary, Francois Marie Charles Fourier who wrote a lot of interesting stuff on some wild socialist ideas) was a mathematician or an engineer. He did teach at France's foremost school of engineering, the Ecole Polytechnique, so he may be claimed as an engineer, but his title was Professor of Analysis so, I suppose, mathematicians may have a good claim as well (and so have the politicians, since Fourier was for a while Governor of Lower Egypt and had various administrative positions in France, both under Napoleon and under the Bourbons). It is certainly true that mathematicians have written an awful lot about Fourier series. One can find books with lemmas, theorems, and proofs heaped high upon each other. I don't think an applied scientist will need much of that. Fourier series are a beautiful example of a subject which can be tackled intuitively, in which it is quite clear whether one has got the right solution or not. Mathematical rigour is not much needed. I shall try to avoid it as much as possible.

My aim is to introduce you to the theory of Fourier series slowly and painlessly with the aid of a large number of examples. I think it is a question of practice. The whole thing might look baffling at the beginning, but after a while, quite imperceptibly, it will become common sense. It should become common sense. If at the end of the lectures any of you feel that the concepts are still unclear, the fault is in the lectures.

1
On approximations

1.1 INTRODUCTION

LET US assume that there is such a thing as truth. Now, if for some reason we can't quite get to the truth we must be satisfied with some approximation. The question then arises, how near is the approximation to the truth? In general, it is impossible to say. There can be millions of possible definitions which all make sense.

As an example, let us take a clock which stopped some years ago at 5.00 p.m. and still displays 5.00. Does this clock give a more accurate representation of time than another one which is always, consistently, half an hour fast? Most men of practical disposition would favour the second clock but, you must agree, it is a matter of definition. The first clock is exactly on time twice a day, whereas the second clock never shows the right time. Mathematically speaking, the question is whether the dashed line or the dotted line is a better approximation to the curve drawn in Fig. 1.1 by continuous lines.

In engineering practice the problem of approximation often appears in the form of having some functional relationship and then trying to approximate it with simple functions.

An example may be seen in Fig. 1.2. For various values of x a measurement is made with the result y yielding the crosses shown. A simple function that will tend to be close to all the crosses is the straight line shown. One may argue how exactly that straight line should be determined and whether the line in Fig. 1.2 is the best approximation we can have, but there is no doubt that our line represents a reasonable approximation.

It also happens quite often that the function to be approximated comes about not as the result of experimental observations but as a mathematical formula and the aim is to approximate it with the aid of some other functions.

1.2 Approximation to a cosine function

As our first example we shall take the function $\cos x$ in the interval $0 < x < \pi/2$, as shown in Fig. 1.3 and we shall try to approximate it

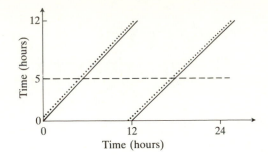

Fig. 1.1 Exact time (——); time shown by clock 1 (– – –); time shown by clock 2 (· · · · ·)

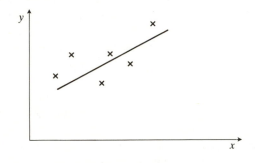

Fig. 1.2 Straight line approximates the curve given by crosses

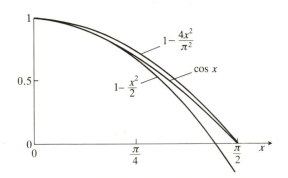

Fig. 1.3 Polynomial approximations to cos x

within that interval with a polynomial. Just to show you the variety of possible approximations, we shall do this in three different ways.

1. *Taylor's series.* I shall not say much about the general rules of how to obtain the various terms of the series, as you are well familiar with them, I just want to mention that in order to find the series around a given point you need to know the value of the function, its first derivative, its second derivative, and so on. The more derivatives you take into account, the better is the approximation. As you know, a quadratic approximation is given as

$$\cos x \cong 1 - \frac{x^2}{2}, \tag{1.1}$$

plotted also in Fig. 1.3. It looks quite good in the vicinity of $x = 0$ but gets worse for larger values of x. At the end of the interval, $x = \pi/2$, the error is 0.2337. It is not negligible.

If we include the quartic term as well in the familiar form

$$\cos x \cong 1 - \frac{x^2}{2!} + \frac{x^4}{4!}, \tag{1.2}$$

the approximation in the given interval is so good that it is within the thickness of the line in Fig. 1.3. Calculating the value of $\cos x$ at $x = \pi/2$ will now yield 0.0199 instead of zero, a much better approximation than before.

2. *Matching at given points.* This is a very simple technique indeed. We demand that the approximating function agrees with the original function at certain points. Let us choose, for example, the quadratic

$$y = a + cx^2 \tag{1.3}$$

to approximate the cosine function, and match the two functions at the points $x = 0$ and $x = \pi/2$. That will yield

$$y(0) = \cos 0 = 1 = a \tag{1.4}$$

and

$$y\left(\frac{\pi}{2}\right) = \cos\frac{\pi}{2} = 0 = a + c\frac{\pi^2}{4}, \tag{1.5}$$

whence

$$a = 1 \quad \text{and} \quad c = -\frac{4}{\pi^2}. \tag{1.6}$$

The approximating function

$$y = 1 - \frac{4x^2}{\pi^2} \tag{1.7}$$

is shown in Fig. 1.3. At $x = \pi/2$ the agreement is now exact but it is poorer around the middle of the interval. At $x = \pi/4$ we find that the error is 0.0429.

If we wish to improve the approximation, we could, of course, opt again for a higher-order polynomial. We may, for example, demand that the function

$$y = a + cx^2 + ex^4 \tag{1.8}$$

agree with the cosine function at the points $x = 0$, $\pi/4$, $\pi/2$, which yields the equations

$$y(0) = \cos 0 = 1 = a; \tag{1.9}$$

$$y\left(\frac{\pi}{4}\right) = \cos\frac{\pi}{4} = \frac{\sqrt{2}}{2} = a + c\frac{\pi^2}{16} + e\frac{\pi^4}{256}; \tag{1.10}$$

$$y\left(\frac{\pi}{2}\right) = \cos\frac{\pi}{2} = 0 = a + c\frac{\pi^2}{4} + e\frac{\pi^4}{16}; \tag{1.11}$$

whence a, c, and e can be determined. The approximating function obtained this way will obviously be superior to the previous quadratic function. I shall not work out the values of the coefficients, it's straightforward enough, but I want to point out one snag with this method. The value of c calculated from eqns (1.9)–(1.11) will be different from that calculated previously from eqns (1.4) and (1.5). This is a nuisance because as we go to higher approximations we have to recalculate the coefficients each time.

3. *Least-squares method*. This is yet another way of doing the approximation. It does not match the two functions at particular points, but rather uses an 'average' criterion: how close the two functions are to each other considering the whole interval.

Let us take our approximating function again in the form

$$y(x) = a + cx^2. \tag{1.12}$$

The difference between the approximating function and our exact function is

$$D(x) = y(x) - \cos x. \tag{1.13}$$

We want this difference to be small. Since it is just as bad for the approximation when the difference is negative it makes sense to consider instead the square of the difference, i.e. the function

$$D^2(x) = [y(x) - \cos x]^2. \tag{1.14}$$

We say now that we have found the optimum approximating function when the average of this difference square over the interval is a minimum. Mathematically,

$$\overline{D^2} = \frac{2}{\pi} \int_0^{\pi/2} D^2(x)\, dx = \text{Min.} \tag{1.15}$$

If we perform the integration by substituting the function $D(x)$ from eqn (1.13), then obviously the result will depend on the coefficients a and c. It is reasonable to assume that by choosing a and c judiciously the value of $\overline{D^2}$ can be minimized.

The rest is simple in principle, though less so in practice because we have to perform a number of integrations. Let us write them out:

$$\overline{D^2} = \frac{2}{\pi} \int_0^{\pi/2} [\cos x - (a + cx^2)]^2\, dx$$

$$= \frac{2}{\pi} \left\{ \int_0^{\pi/2} \cos^2 x\, dx - 2a \int_0^{\pi/2} \cos x\, dx - 2c \int_0^{\pi/2} x^2 \cos x\, dx \right.$$

$$\left. + a^2 \int_0^{\pi/2} dx + 2ac \int_0^{\pi/2} x^2\, dx + c^2 \int_0^{\pi/2} x^4\, dx \right\}. \tag{1.16}$$

By introducing the notations

$$A = \frac{2}{\pi} \int_0^{\pi/2} \cos^2 x\, dx, \qquad B = \frac{2}{\pi} \int_0^{\pi/2} \cos\, dx, \qquad C = \frac{2}{\pi} \int x^2 \cos x\, dx,$$

$$E = \frac{2}{\pi} \int_0^{\pi/2} dx, \qquad F = \frac{2}{\pi} \int_0^{\pi/2} x^2\, dx, \qquad G = \frac{2}{\pi} \int_0^{\pi/2} x^4\, dx, \tag{1.17}$$

the above equation reduces to

$$\overline{D^2} = A - 2aB - 2cC + a^2E + 2acF + c^2G. \tag{1.18}$$

We want to minimize this function. But minimize it as a function of what? Note that A, B, C, E, F, G are now constants; our variables are a and c. Equation (1.18) represents a function in two variables. To find its minimum we need to take the partial derivatives with respect to both a and c. We obtain for the minimum condition

$$\frac{\partial \overline{D^2}}{\partial a} = -2B + 2aE + 2cF = 0, \tag{1.19}$$

$$\frac{\partial \overline{D^2}}{\partial c} = -2C + 2aF + 2cG = 0, \tag{1.20}$$

whence the optimum values of a and c may be determined. The calculation itself is a bit lengthy but straightforward. Having found a and c we can write the polynomial as

$$y(x) = 0.9801 - 0.4176x^2. \tag{1.21}$$

This function lies quite close to cos x. I shall not plot it, because I don't want to make Fig. 1.3 too crowded, but I shall give the error at a few places. It is 0.0199 at $x = 0$, 0.0150 at $x = \pi/4$, and 0.0502 at $x = \pi/2$. Note that we now have an error at $x = 0$, which we did not have with either of the previous methods.

1.3 Periodic functions

We are now getting closer to the subject we wish to study. Are there some special considerations when we want to approximate a periodic function? We know that it is possible to approximate a periodic function with the sum of non-periodic functions, as shown for example by the Taylor series expansion of a sine function into a polynomial. But that is not the only way of doing it. In the next two sections we shall be concerned with some qualitative examples introducing the concept of the Fourier series, its advantages and disadvantages. But before talking about what a Fourier series can do, let me say a few words about periodic functions in general.

The best-known examples are provided by Nature. Most of you will be familiar with the consistency with which day follows night and summer follows winter. (It was this periodic relationship which was pointed out by Shelley when he wrote in a somewhat different context that "if winter comes can spring be far behind?"). Thus if you measured the temperature, for example, as a function of time you would certainly find periodic variations reflecting the ways our planet moves relative to the Sun.

Another well-known example of a periodic function is a wave on the surface of water. It shows a nice periodic pattern. Sound waves in solids, electromagnetic waves in vacuum, and the voltage in the socket on the wall are further examples of periodic functions which we know of but have never had the chance to see.

I think it is in general an advantage to attach some physics to it when you visualize a periodic function, but that is not obligatory. If you are happy with mathematics on its own, that's fine, for quite some time you may rely on mathematics alone.

1.4 The claim

Ar first hearing, the claim sounds preposterous. It is claimed that if you take a periodic function (e.g. the one shown in Fig. 1.4a, where for simplicity the period is taken as 2π) and then take the fundamental cosine component (Fig. 1.4b), the fundamental sine component (Fig.

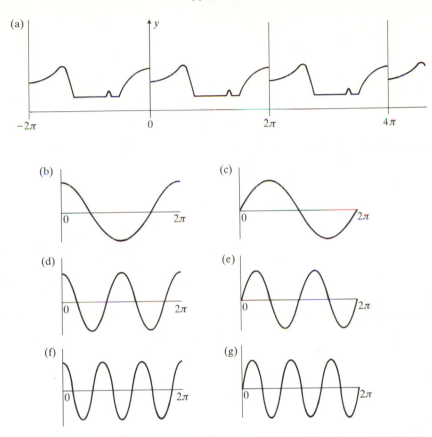

Fig. 1.4 (a) An arbitrary periodic function; (b)–(g) cosine and sine functions having periods of 2π, π, and $2\pi/3$

1.4c), the second harmonic cosine component (Fig. 1.4d), the second harmonic sine component (Fig. 1.4e), the third harmonic cosine component (Fig. 1.4f), the third harmonic sine component (Fig. 1.4g), and so on to infinity, and if you take the right amount of each component and add them up, and finally add to this sum a judiciously chosen constant, you will get the curve shown in Fig. 1.4a.

The reason why the claim sounds so ridiculous is because it is plainly against common sense. One may argue that in order to approximate $f(x)$ between x_1 and x_2 (see Fig. 1.5a) one needs part of a sinusoidal (as shown in Fig. 1.5b by solid lines) but surely not the rest of the function (shown by dotted lines) because adding that will only make the approximation worse outside the x_1-x_2 interval.

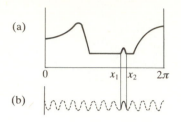

Fig. 1.5 (a) The periodic function of Fig. 1.4a. (b) An approximating function in the range x_1, x_2.

I would like to assure you, however, that the claim is in some sense correct. By adding more and more harmonics in the right proportion an arbitrary periodic function can indeed be approximated with any desired accuracy. True, mathematicians do have some reservations about the arbitrariness of the function, but the theorem is valid for any function that will ever come up in the practice of an applied scientist. The proof of convergence may be found in most textbooks. If you are fond of proofs do look it up by all means, but I do not think it is worth including it in the present course. My intention is to give lots of examples (more than given usually), adhering to the simple principle that seeing is believing and that seeing lots of examples engenders genuine belief.

1.5   A few examples showing the build-up

I shall give now a few numerical examples showing how a periodic function is approximated by its Fourier series.

For our first example the periodic function to be approximated is shown in Fig. 1.6. It is a simple kind of function; it takes the value of 1 between $-\pi/4$ and $\pi/4$, and it takes zero outside this region. The period is 2π. Can we approximate this function by a Fourier series? Yes, we can. I shall give here a blow-by-blow account.

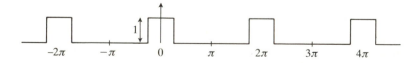

Fig. 1.6 A series of rectangular pulses

First we shall have to choose one period of this periodic function. The period must of course be 2π but the function need not be chosen between 0 and 2π. For symmetry reasons we shall take the interval between $-\pi$ and π, so that we have the even function shown in Fig. 1.7a. It then makes sense that an even function is built up from cosine components and all the sine components are zero.

The Fourier series may then be written in the form

$$f(x) = \frac{a_0}{2} + \sum_{k=1}^{\infty} a_k \cos kx, \tag{1.22}$$

where $f(x)$ is plotted in Fig. 1.7a. Don't worry about the constant, it will be clear later why it is written in that particular form.

How can we find a_k, the amplitude of the kth component? This is of course the relevant question, the 64,000 dollar question as the Americans would say. I shall show you the technique later, but first I would like to convince you that we can indeed approximate a periodic function by adding up lots of individual components. So just trust for the time being that I managed to determine the amplitudes correctly.

I shall now denote the kth component by

$$p_k = a_k \cos kx \tag{1.23}$$

and shall define the function

$$f_n(x) = \sum_{k=1}^{n} a_k \cos kx = \sum_{k=1}^{n} p_k, \tag{1.24}$$

i.e. $f_n(x)$ is the approximate function we obtain having added up n components.

So let us look at (Fig. 1.7b) $f_1 = p_1$, which is the fundamental component having a period of 2π. We have to admit that it does not much resemble Fig. 1.7a. Let's then add the second harmonic (which has a period of π) shown in Fig. 1.7c. The result is f_2, which may be seen in Fig. 1.7d. There is still no resemblance. Let us then add the third harmonic (Fig. 1.7e), resulting in f_3. If we look at Fig. 1.7b, d, and f we see that the middle portion of the function gets narrower. Is this good? Not really. We don't want a spike, we need a square kind of thing. But the step from f_2 to f_3 should rather be described as *reculer pour mieux sauter*. The lobe in the middle does indeed get narrower, but when the fifth harmonic (Fig. 1.7g) is added then the middle lobe widens again as may be seen in Fig. 1.7h.

Have I forgotten to add the fourth harmonic? No, I haven't. The symmetry properties of this periodic function are such that all the $4n$ harmonics (where n is an integer) have zero amplitude.

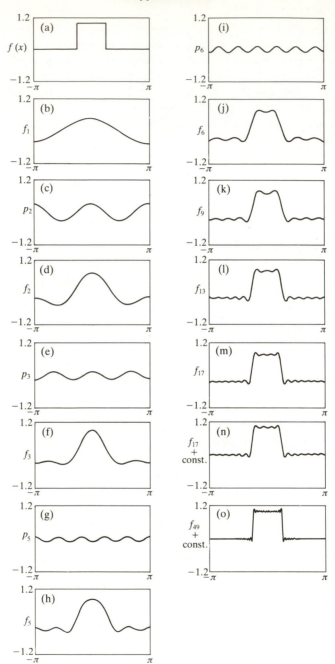

Fig. 1.7 (a) A periodic function defined between $-\pi/2$ and $\pi/2$ ($f(x) = 1$ for $|x| < \pi/4$ and zero otherwise) to be approximated by a Fourier series. For (b) to (o) note that f_n denotes the sum of the first n harmonics excluding the constant term and p_n denotes the nth harmonic

Let us now add the sixth harmonic (Fig. 1.7i), which results in f_6 shown in Fig. 1.7j. It should now be clear what these higher harmonics do. They add and subtract from the function just at the right places. Figure 1.7j looks definitely more squarish. Now, with a little imagination one can perhaps believe that when we add more and more components we shall be able to get close to the asymmetric square wave of Fig. 1.7a.

I shall not show any more of the individual components but only their sums. With 9 harmonics we get Fig. 1.7k: with 13 harmonics we get Fig. 1.7l: and with 17 harmonics we get Fig. 1.7m. A pattern is now clearly visible. Every time we go up by four harmonics we have one more ripple in the main lobe and it always gets a little closer to the square wave.

One thing is of course still missing. We have not so far considered the constant term. Adding it to the curve in Fig. 1.7m we obtain Fig. 1.7n, which is a good approximation to Fig. 1.7a.

How good is the approximation? How can we express it in words? We can say that by adding more harmonics the ripples become smaller and they will of course vary faster. It is not unreasonable to expect at this stage that as the frequency goes to infinity the amplitude of the ripple goes to zero.

Let us now jump to the curve given by the first 49 harmonics (Fig. 1.7o) to see whether we guessed the tendency right. It is true indeed that the frequency of the ripples gets higher and their amplitude gets smaller. In the vicinity of the sudden transition from 1 to 0, however, the situation fails to improve, the overshoot on one side and the undershoot on the other side show no diminution. This is unfortunately a real effect (known as the Gibbs phenomenon—more about it later) showing that not even the Fourier series is perfect. I hope you will, none the less, agree that to have such a good approximation by adding up lots of little functions must border on the miraculous. The convergence (as measured by the number of terms added) may be slow, but that is a characteristic of the example chosen. I have deliberately chosen an example in which the Fourier series does not appear in its best colours. I think it is a good psychological approach never to overpraise a product (however remarkable its properties are) because customers can easily get disenchanted when it does not live up to all the (usually inflated) expectations.

For our second example, let us choose a function (Fig. 1.8a) which is not very far away from a sinusoidal. It is defined in the $-\pi < x < \pi$ region as follows:

$$f(x) = \begin{cases} \sin x & |x| < \pi/4 \\ -\sqrt{2}/2 & -3\pi/4 < x < -\pi/4 \\ \sqrt{2}/2 & \pi/4 < x < 3\pi/4 \\ \sin x & 3\pi/4 < |x| < \pi \end{cases} \quad \text{if} \qquad (1.25)$$

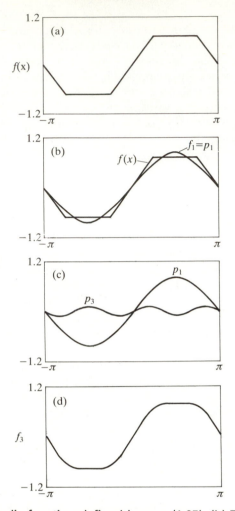

Fig. 1.8 (a) A periodic function defined by eqn (1.25). (b) The function and its approximation by the fundamental component. (c) The fundamental component and the third harmonic. (d) The sum of the first three harmonics

We obtain this function by chopping off the top of a sinusoidal (the signal goes through a limiter, one would say in electrical engineering terms). So we may expect that the fundamental component will already give a good approximation. This is indeed so, as may be seen in Fig. 1.8b where both $f(x)$ and the fundamental component p_1 are plotted. Note that the amplitude of the fundamental component is smaller than that of the original sinusoidal. If one determines its value properly, it comes to 0.725.

Owing to symmetry reasons, there is no second harmonic (believe it for the moment, it will be proved later) but there is a third harmonic p_3, as may be seen in Fig. 1.8c where both the third harmonic and the fundamental are plotted. If we add the two together we obtain the function f_3 shown in Fig. 1.8d. Note that we needed to add only two sinusoidals and we obtained a remarkably good approximation to the chopped sinusoidal shown in Fig. 1.8a.

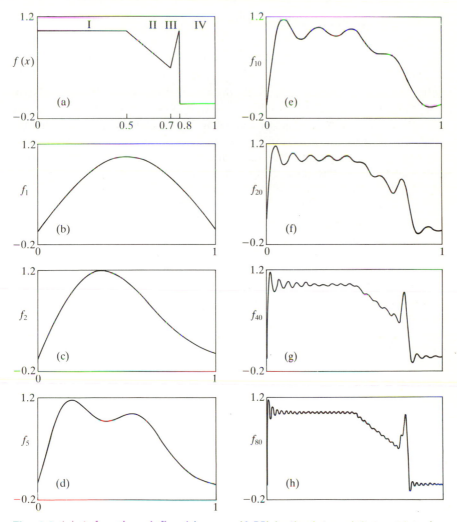

Fig. 1.9 (a) A function defined by eqn (1.26) in the interval $0 < x < 1$ to be approximated by sine functions. For (b) to (h) note that f_n denotes the sum of the first N harmonics. There is no constant term

For our third example, let us choose something outrageous like the function shown in Fig. 1.9a which is, actually, still better looking than the one in Fig. 1.4a. It is defined as follows

$$f(x) = \begin{cases} 1 & 0 < x < 0.5 & \text{region I} \\ 2(1-x) & 0.5 < x < 0.75 & \text{region II} \\ 10x - 7 & 0.75 < x < 0.8 & \text{region III} \\ 0 & 0.8 < x < 1 & \text{region IV} \end{cases} \quad \text{if} \quad (1.26)$$

I shall approximate it in the form

$$f_N(x) = \sum_{k=1}^{N} b_k \sin k\pi x \qquad (1.27)$$

with judiciously chosen b_k; f_1 is shown in Fig. 1.9b. It looks fairly non-committal; there are no hints what it will develop into. There is though a slight hint of the right kind of asymmetry in f_2 shown in Fig. 1.9c. Let us now jump to f_5 in Fig. 1.9d. It may be seen that it tries hard to make the beginning of the function rise sharply and also to approximate the flat part of the function in region I. Doubling the number of components to $N = 10$ we can see (Fig. 1.9e) intimations of regions I, II and IV but not yet of region III. Obviously, the faster the function to be approximated varies, the higher are the harmonics necessary for good approximation. Doubling again to $N = 20$, region III may be seen to form in Fig. 1.9f. At $N = 40$ (Fig. 1.9g) all four regions can be clearly distinguished. If we increase N further, the ripples will reduce in amplitude and will have higher frequencies. I shall show just one more curve, for $N = 80$ (Fig. 1.9h). The Gibbs phenomenon, as may now be expected, persists at the sharp changes at $x = 0$ and $x = 0.8$, but otherwise f_{80} may be said to be an excellent approximation of $f(x)$ shown in Fig. 1.9a.

2
Finding the Fourier series

As YOU have probably suspected, in a course of mathematics one is bound, sooner or later, to do a little mathematics. We shall now find the formulae for the coefficients of the expansion.

2.1 Derivation of the coefficients

In the general case one has both cosine and sine terms, which I shall include below, but I shall for the time being restrict the generality and shall assume that the period of the periodic function is 2π. The Fourier series may then be written in the form

$$f(x) = \frac{a_0}{2} + \sum_{k=1}^{\infty} a_k \cos kx + \sum_{k=1}^{\infty} b_k \sin kx. \qquad (2.1)$$

The problem is to find all the coefficients. It seems a formidable problem. The normal expectation is that in order to find N coefficients one needs to set up N simultaneous equations. This is what we did when we found the coefficients of various polynomials in Section 1.2. It turns out that in the present case simultaneous equations are not needed at all. By relying on certain properties of the sine and cosine functions, each coefficient may be determined separately.

Let us start by finding the term $a_0/2$. We may find it by taking the average of both the left-hand side and of the right-hand side in eqn (2.1). How do we take the average of a function? We integrate it, and then divide by the length of the interval in which we look for the average. Our interval is the period 2π; thus we end up with

$$\frac{1}{2\pi} \int_0^{2\pi} f(x)\, dx = \frac{1}{2\pi} \int_0^{2\pi} \frac{a_0}{2}\, dx$$

$$+ \sum_{k=1}^{\infty} a_k \frac{1}{2\pi} \int_0^{2\pi} \cos kx\, dx + \sum_{k=1}^{\infty} b_k \frac{1}{2\pi} \int_0^{2\pi} \sin kx\, dx. \quad (2.2)$$

Let us look at the integral

$$\int_0^{2\pi} \cos kx \, dx. \tag{2.3}$$

The integrand is $\cos kx$, which has exactly k periods between 0 and 2π. We know that the operation of integration within two limits gives the area under the curve. But having exact number of periods within our interval means that we have the same area above the axis as below the axis, and therefore the integration shown in eqn (2.3) yields zero.

There is no need, of course, to appeal to graphical concepts: we can get the same result by actually doing the integration,

$$\int_0^{2\pi} \cos kx \, dx = \frac{1}{k} \sin kx \, \bigg|_0^{2\pi} = 0, \tag{2.4}$$

and of course the same applies to the integration of $\sin kx$: that integrates out to zero as well. There is one more integration to do:

$$\frac{1}{2\pi} \int_0^{2\pi} \frac{a_0}{2} \, dx = \frac{a_0}{2}, \tag{2.5}$$

and we are left with

$$\frac{a_0}{2} = \frac{1}{2\pi} \int_0^{2\pi} f(x) \, dx, \tag{2.6}$$

where $f(x)$ is a known function. In order to find $a_0/2$, all we need to do is to take its average value, i.e. integrate it for the interval 0 to 2π and divide the result by 2π. It is as simple as that.

In order to derive a_k I shall now multiply eqn (2.1) by $\cos lx$ (where l is an integer) and integrate again both sides of the equation for the interval 0 to 2π. We obtain

$$\int_0^{2\pi} f(x) \cos lx \, dx = \frac{a_0}{2} \int_0^{2\pi} \cos lx \, dx$$

$$+ \sum_{k=1}^{\infty} a_k \int_0^{2\pi} \cos kx \cos lx \, dx + \sum_{k=1}^{\infty} b_k \int_0^{2\pi} \sin kx \cos lx \, dx. \tag{2.7}$$

Before we perform the integrations, let us write down some well-known trigonometric formulae:

$$\cos kx \cos lx = \tfrac{1}{2}[\cos(k+l)x + \cos(k-l)x]; \tag{2.8}$$

$$\sin kx \sin lx = \tfrac{1}{2}[\cos(k-l)x - \cos(k+l)x]; \tag{2.9}$$

$$\sin kx \cos lx = \tfrac{1}{2}[\sin(k+l)x + \sin(k-l)x]. \tag{2.10}$$

Using the above formulae, we find

$$\int_0^{2\pi} \sin kx \cos lx \; dx = \frac{1}{2} \left\{ \int_0^{2\pi} \sin(l+k)x \; dx + \int_0^{2\pi} \sin(k-l)x \; dx \right\} = 0$$

$$(2.11)$$

and

$$\int_0^{2\pi} \cos kx \cos lx \; dx = \frac{1}{2} \left\{ \int_0^{2\pi} \cos(k+l)x \; dx + \int_0^{2\pi} \cos(k-l)x \; dx \right\} = 0.$$

$$(2.12)$$

This is correct, isn't it? Well, nearly. We have overlooked one little case when $l = k$. In that case $\cos(k-l)x = 1$ and the integration yields

$$\int_0^{2\pi} \cos^2 kx \; dx = \frac{1}{2} \int_0^{2\pi} (1 + \cos 2kx) \; dx = \frac{1}{2} \int_0^{2\pi} dx = \pi. \qquad (2.13)$$

So we get the interesting result that

$$\int_0^{2\pi} \cos kx \cos lx \; dx = \begin{matrix} 0 \\ \pi \end{matrix} \quad \text{if} \quad \begin{matrix} k \neq l \\ k = l \end{matrix} \qquad (2.14)$$

The relationships given in eqns (2.11) and (2.14) are known as orthogonality relationships. When the integrations yield zero, the two functions are said to be orthogonal† to each other.

Let us not lose the sight of the aim. Our intention is to perform the integrations in eqn (2.7). Out of the summation

$$\sum_{k=1}^{\infty} b_k \int_0^{2\pi} \cos kx \sin lx \; dx \qquad (2.15)$$

nothing remains, because the integral, always, without exception, yields zero. The other summation

$$\sum_{k=1}^{\infty} a_k \int_0^{2\pi} \cos kx \cos lx \; dx, \qquad (2.16)$$

† Why orthogonal? What does orthogonal mean? Well, the nearest analogy is with vectors, with ordinary three-dimensional vectors. You will recall the scalar product of two such vectors and in particular that it is zero when the two vectors are orthogonal (perpendicular) to each other.

Just as one can think of a set of three-dimensional vectors so one should be able to imagine the set of functions $\cos kx$ (one function for each value of k). Just as one can define the operation of scalar product between two vectors, it is also possible to define an operation between the elements of our set of functions. We shall now define such an operation. We shall take two functions from the set (we may take the same function twice), multiply them together and integrate them between the limits of 0 and 2π. With this definition we have now ordered a scalar to every pair of functions chosen from the set. If this scalar happens to be zero when the two functions are different, we say the two functions are orthogonal to each other.

however, as we have just seen, does not completely vanish. One term in the summation survives when $k = l$, yielding πa_l.

The first integration of the right-hand side of eqn (2.7) will obviously give zero, hence the whole thing reduces to

$$\int_0^{2\pi} f(x) \cos lx \, dx = \pi a_l \tag{2.17}$$

or

$$a_l = \frac{1}{\pi} \int_0^{2\pi} f(x) \cos lx \, dx. \tag{2.18}$$

But l can be any integer. Thus we have succeeded in finding all the coefficients of the set of cosine functions.

As you may guess, we can find b_l by similar means. Multiply eqn (2.1) by $\sin lx$ and integrate over the interval 0 to 2π. We get

$$\int_0^{2\pi} f(x) \sin lx \, dx = \frac{a_0}{2} \int_0^{2\pi} \sin lx \, dx + \sum_{k=1}^{\infty} a_k \int_0^{2\pi} \cos kx \cos lx \, dx$$

$$+ \sum_{k=1}^{\infty} b_k \int_0^{2\pi} \sin kx \sin lx \, dx. \tag{2.19}$$

We have already proved that the second integral on the right-hand side vanishes. The first integral may also be seen to vanish, so we are left with the third integral, which yields

$$\int_0^{2\pi} \sin kx \sin lx \, dx = \frac{1}{2} \left\{ \int_0^{2\pi} \cos(k-l)x \, dx - \int_0^{2\pi} \cos(k+l)x \, dx \right\}$$

$$= \begin{matrix} 0 \\ \pi \end{matrix} \quad \text{if} \quad \begin{matrix} k \neq l \\ k = l \end{matrix}. \tag{2.20}$$

Rearranging eqn (2.19), we obtain for the coefficient of the sine functions

$$b_l = \frac{1}{\pi} \int_0^{2\pi} f(x) \sin lx \, dx. \tag{2.21}$$

We have reached the end of the road. We have succeeded in determining all the coefficients in eqn (2.1): a_0 from eqn (2.6), a_k from eqn (2.18), and b_k from eqn (2.21). By the way, it should now be clear why the constant term is usually written in the form $a_0/2$. In that case eqn (2.6) is superfluous and a_0 can be calculated from eqn (2.18).

Finally, I must state the theorem with which I probably should have started concerning the function $f(x)$. What are the conditions that $f(x)$ should satisfy in order to have a Fourier series? There are various ways of

giving sufficient conditions. I shall give the following two conditions:
(i) $f(x)$ should have a finite number of maxima, minima and discontinuities in one period;
(ii) the integral

$$\int_{-\pi}^{\pi} |f(x)|\, dx \qquad (2.22)$$

should be convergent.

If the above conditions are satisfied then the Fourier series converges to $f(x)$ as the number of terms tends to infinity.

These conditions should not worry you. In engineering practice they are invariably satisfied (it might just happen that owing to certain approximations in the model the integral in eqn (2.22) will diverge, but then look again at your approximations).

We still need to know what happens at discontinuities. How can the Fourier series make up its mind to what point to converge? Will it converge to the lower value or will it converge to the upper value? The common-sense expectation is that it will converge to the average value in between. And indeed this is what the theorem says.

For all values of x the function $f(x)$ converges to the value

$$\tfrac{1}{2}[f(x+0)+f(x-0)], \qquad (2.23)$$

where

$$f(x+0) = \lim_{\substack{h \to 0 \\ h>0}} f(x+h) \quad \text{and} \quad f(x-0) = \lim_{\substack{h \to 0 \\ h>0}} f(x-h). \quad (2.24)$$

2.2   Symmetry considerations

Let us recall what we mean by even and odd functions. For an even function

$$f(x) = f(-x) \qquad (2.25)$$

and for an odd function

$$f(x) = -f(-x). \qquad (2.26)$$

Secondly, we need to know that when we take the product of two even functions we have

$$\text{even} \times \text{even} = \text{even}, \qquad (2.27)$$

$$\text{even} \times \text{odd} = \text{odd}, \qquad (2.28)$$

$$\text{odd} \times \text{odd} = \text{even}. \qquad (2.29)$$

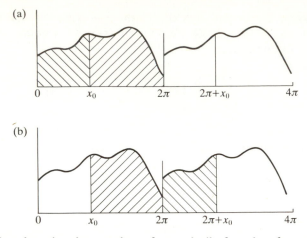

Fig. 2.1 Showing that integration of a periodic function for one period is independent of the starting point

In the third place I want to mention the rules that between symmetric limits c,

$$\int_{-c}^{c} (\text{odd function}) \, dx = 0 \qquad (2.30)$$

$$\int_{-c}^{c} (\text{even function}) \, dx = 2 \int_{0}^{c} (\text{even function}) \, dx \qquad (2.31)$$

both of which are obvious if you think of the area under the curves.

In the fourth place I want to use the theorem that when a periodic function is integrated for one period the result is independent of the starting point of the integration. This may be clearly seen in Fig. 2.1a and b, where integrations from 0 to 2π and from x_0 to $x_0 + 2\pi$ are shown. In both cases we have the integration from x_0 to 2π so that part of the area (stippled) is identical. The difference is that in the first case we integrate from 0 to x_0, whereas in the second case we integrate from 2π to $2\pi + x_0$. However, since the function is periodic it has the same values from 0 to x_0 as from 2π to $2\pi + x_0$ and therefore the areas under the curves (hatched) must also be identical.

Armed with all these theorems we can now work out some simplified formulae for the Fourier coefficients when $f(x)$ is either even or odd. We shall use eqns (2.18) and (2.21) but take the period from $-\pi$ to π.

When $f(x)$ is even we have

$$a_k = \frac{1}{\pi} \int_{-\pi}^{\pi} f(x) \cos kx \, dx. \qquad (2.32)$$

But both $f(x)$ and $\cos kx$ are even functions, therefore eqn (2.32) reduces to

$$a_k = \frac{2}{\pi} \int_0^\pi f(x) \cos kx \, dx. \tag{2.33}$$

This is not much of a simplification, but we fare better with b_k. We find

$$b_k = \frac{1}{\pi} \int_{-\pi}^\pi f(x) \sin kx \, dx = 0 \tag{2.34}$$

because even × odd = odd, and the integral of an odd function between symmetric limits is zero.

When $f(x)$ is odd then similar considerations lead to

$$a_k = 0 \qquad \text{and} \qquad b_k = \frac{2}{\pi} \int_0^\pi f(x) \sin kx \, dx. \tag{2.35}$$

So it is always worth investigating whether a function is even or odd. It leads to a lot of saving in labour. Just one more word about the constant term. It needs to be disregarded when the function is tested for 'oddness' (or is it called 'oddity'?).

Can we have further simplifications on account of symmetry? The answer is yes. Some coefficients will be absent if the function to be expanded has some additional symmetry properties.

Take, for example, the case when $f(x)$ is an even function, so that a_k can be determined from eqn (2.32), and assume further that $f(x)$ is odd around $\pi/2$ by this I mean that it satisfies the condition $f(x) = -f(\pi - x)$. An example of such a function is shown in Fig. 2.2a.

Let us now work out the even harmonics for this case. They are given by

$$a_{2k} = \frac{2}{\pi} \int_0^\pi f(x) \cos 2kx \, dx. \tag{2.36}$$

Using the properties of the function mentioned above, we can split the integral into two parts, leading to

$$a_{2k} = \frac{2}{\pi} \left\{ \int_0^{\pi/2} f(x) \cos 2kx \, dx - \int_{\pi/2}^\pi f(\pi - x) \cos 2kx \, dx \right\}. \tag{2.37}$$

We may now introduce the new variable $s = \pi - x$ in the second integral. Transforming the limits and noting that $\cos 2k(\pi - s) = \cos 2ks$, we find that $a_{2k} = 0$.

Summarizing, if $f(x)$ is even about the origin and, in addition, odd about $\pi/2$, then there are no sine components and also the even cosine components are missing.

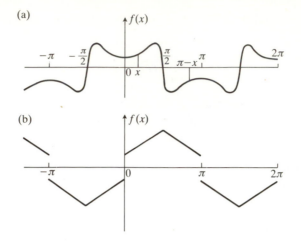

Fig. 2.2 (a) $f(x)$ is even about 0 and odd about $\pi/2$; (b) $f(x)$ is odd about 0 and even about $\pi/2$

Similarly, when $f(x)$ is odd about the origin and, in addition, even about $\pi/2$ (see Fig. 2.2b for an example of such a function), then there are no cosine components and also the even sine components are missing.

As my last example on symmetry let me recall the function shown in Fig. 1.7a. When I built it up from its Fourier components I said that the 4th, 8th, 12th, etc., harmonics were zero. Now we can prove this easily. The function is even, so that eqn (2.32) applies. We obtain

$$a_{4k} = \frac{2}{\pi} \int_0^{\pi/4} \cos 4kx \, dx = \frac{2}{\pi} \frac{1}{4k} \sin 4kx \Big|_0^{\pi/4} = 0. \qquad (2.38)$$

2.3 A few worked examples

To be able to use Fourier series one needs to acquire a certain 'feel' for the subject and in order to have the 'feel' one must work through lots of examples. I shall try to do that here.

Example I. The simplest example that you can find in practically every textbook is the square wave function shown in Fig. 2.3. It is a good function to start with because it needs the minimum amount of mathematical labour. It is defined in the interval $-\pi < x < \pi$ as

$$f_I(x) = \begin{matrix} -1 \\ 1 \end{matrix} \quad \text{if} \quad \begin{matrix} -\pi < x < 0 \\ 0 < x < \pi, \end{matrix} \qquad (2.39)$$

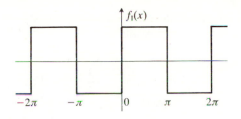

Fig. 2.3 A square wave function

and outside this interval it is continued with the relationship

$$f_I(x + 2\pi) = f_I(x),\qquad(2.40)$$

which makes it a periodic function of period 2π.

It is easy to see that $f_I(x)$ as defined above is an odd function; thus the cosine components are zero and the sine components can be determined from eqn (2.35), yielding

$$b_k = \frac{2}{\pi}\int_0^\pi \sin kx\,dx = \frac{2}{\pi}\left(-\frac{1}{k}\right)\cos kx\,\bigg|_0^\pi = -\frac{2}{k\pi}[(-1)^k - 1].\quad(2.41)$$

Note that $f_I(x)$ is even about $x = \pi/2$; hence the even sine components are missing, as discussed in the previous section. The corresponding Fourier series is

$$f_I(x) = \frac{4}{\pi}\left(\sin x + \tfrac{1}{3}\sin 3x + \tfrac{1}{5}\sin 5x + \cdots\right) = \frac{4}{\pi}\sum_{i=1}^{\infty}\frac{\sin(2i-1)x}{2i-1}\quad(2.42)$$

Example II. We shall follow the fashion and choose for our next example the function which ranks next in popularity, namely the triangular function (Fig. 2.4) defined as

$$f_{II}(x) = \begin{array}{l} -x \\ x \end{array}\quad\text{if}\quad \begin{array}{l} -\pi < x < 0 \\ 0 < x < \pi, \end{array}\qquad(2.43)$$

and continued periodically outside the $-\pi$ to π interval.

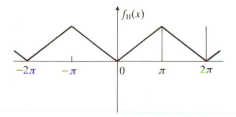

Fig. 2.4 A triangular function

This is an even function; thus the sine components are missing and the coefficients of the cosine components may be obtained from eqn (2.33) as follows:

$$a_k = \frac{2}{\pi} \int_0^\pi x \cos kx \, dx = \frac{2}{\pi} \left\{ \frac{1}{k} x \sin kx \, \Big|_0^\pi - \frac{1}{k} \int_0^\pi \sin kx \, dx \right\}$$

$$= \left(-\frac{2}{k\pi} \right) \left(-\frac{1}{k} \right) \cos kx \, \Big|_0^\pi = \frac{2}{k^2\pi} [(-1)^k - 1], \qquad k \neq 0. \quad (2.44)$$

Note again that the function is odd about $x = \pi/2$ (having disregarded the constant component) and therefore $a_{2k} = 0$. Further,

$$a_0 = \frac{2}{\pi} \int_0^\pi x \, dx = \pi, \qquad (2.45)$$

leading to the Fourier series

$$f_\text{II}(x) = \frac{\pi}{2} - \frac{4}{\pi} \sum_{i=1}^\infty \frac{\cos(2i-1)x}{(2i-1)^2} \qquad (2.46)$$

Example III. Let us next define a function in the interval 0 to 2π as follows:

$$f_\text{III}(x) = \frac{x}{2\pi}, \qquad (2.47)$$

and, again, the function is continued periodically outside this interval (Fig. 2.5a).

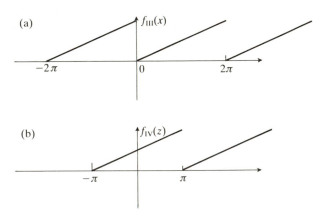

Fig. 2.5 (a) A sawtooth function. (b) The same function in a coordinate system shifted by π

The coefficients may be worked out from the formulae

$$a_k = \frac{1}{\pi} \int_0^{2\pi} \frac{x}{2\pi} \cos kx \, dx = \frac{1}{2\pi^2} \left\{ \frac{1}{k} x \sin kx \Big|_0^{2\pi} - \frac{1}{k} \int_0^{2\pi} \sin kx \, dx \right\} = 0$$

(2.48)

and

$$a_0 = \frac{1}{2\pi^2} \int_0^{2\pi} x \, dx = 1.$$

(2.49)

Further,

$$b_k = \frac{1}{\pi} \int_0^{2\pi} \frac{x}{2\pi} \sin kx \, dx$$

$$= \frac{1}{2\pi^2} \left\{ -\frac{1}{k} x \cos kx \Big|_0^{2\pi} + \frac{1}{k} \int_0^{2\pi} \cos kx \, dx \right\} = -\frac{1}{k\pi}.$$

(2.50)

The Fourier series is

$$f_{\text{III}}(x) = \frac{1}{2} - \frac{1}{\pi} \sum_{k=1}^{\infty} \frac{1}{k} \sin kx$$

(2.51)

Example IV. We shall now ask the question whether the normal rules of shifting a coordinate system still apply to a Fourier series? As an example we introduce the new coordinate z by the transformation

$$z = x - \pi,$$

(2.52)

in which case the function shown in Fig. 2.5a will be transformed to that shown in Fig. 2.5b.

We find the new Fourier series by substituting eqn (2.52) into eqn (2.51), yielding

$$f_{\text{IV}}(z) = \frac{1}{2} - \frac{1}{\pi} \sum_{k=1}^{\infty} \frac{1}{k} \sin k(2 + \pi)$$

$$= \frac{1}{2} - \frac{1}{\pi} \sum_{k=1}^{\infty} \frac{(-1)^k}{k} \sin kz.$$

(2.53)

Let us now find the series by working out the coefficients for the new function. we obtain

$$b_k = \frac{1}{\pi} \int \frac{1}{2\pi} (2 + \pi) \sin kz \, dz = \frac{1}{2\pi^2} \left\{ \int_{-\pi}^{\pi} \sin kz \, dz + \pi \int_{-\pi}^{\pi} \sin kz \, dz \right\}$$

$$= \frac{1}{\pi^2} \left\{ -\frac{1}{k} z \cos z \Big|_0^{\pi} + \frac{1}{k} \int_0^{\pi} \cos kz \, dz \right\} = -\frac{(-1)^k}{k\pi},$$

(2.54)

leading, unsurprisingly, to the Fourier series already given by eqn (2.53).

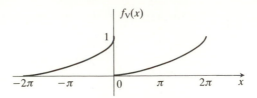

Fig. 2.6 The function $(x/2\pi)^2$ defined in the interval 0 to 2π and extended periodically

Example V. Let us find now the Fourier series of the function shown in Fig. 2.6, defined in the interval 0 to 2π as

$$f_V(x) = \left(\frac{x}{2\pi}\right)^2 . \tag{2.55}$$

The coefficients of the cosine functions may be obtained from

$$a_k = \frac{1}{\pi} \int_0^{2\pi} \left(\frac{x}{2\pi}\right)^2 \cos kx \, dx = \frac{1}{4\pi^3} \left\{ \frac{1}{k} x^2 \sin kx \Big|_0^{2\pi} - \frac{1}{k} \int_0^{2\pi} 2x \sin kx \, dx \right\}$$

$$= -\frac{1}{2k\pi^3} \left\{ -\frac{1}{k} x \cos kx \Big|_0^{2\pi} + \frac{1}{k} \int_0^{2\pi} \cos kx \, dx \right\} = \frac{1}{k^2\pi^2}, \qquad k \neq 0,$$

$$\tag{2.56}$$

whereas

$$a_0 = \frac{1}{4\pi^3} \int_0^{2\pi} x^2 \, dx = \tfrac{2}{3} \tag{2.57}$$

and

$$b_k = \frac{1}{\pi} \int_0^{2\pi} \left(\frac{x}{2}\right)^2 \sin kx \, dx = \frac{1}{4\pi^3} \left\{ -\frac{x^2}{k} \cos kx \Big|_0^{2\pi} + \frac{1}{k} \int_0^{2\pi} 2x \cos kx \, dx \right\}$$

$$= \frac{1}{4\pi^3} \left\{ -\frac{4\pi^2}{k} + \frac{2}{k} \left[\frac{x}{k} \sin kx \Big|_0^{2\pi} - \frac{1}{k} \int_0^{2\pi} \sin kx \, dx \right] \right\} = -\frac{1}{k\pi} . \tag{2.58}$$

The Fourier series is

$$f_V(x) = \frac{1}{3} + \frac{1}{\pi^2} \sum_{k=1}^{\infty} \left[\frac{1}{k^2} \cos kx - \frac{\pi}{k} \sin kx \right] . \tag{2.59}$$

Example VI. With our last example in this section we shall venture into the realm of exponential functions and define our periodic function

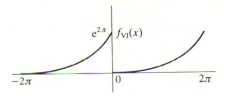

Fig. 2.7 The function e^x defined in the interval 0 to 2π and extended periodically

(Fig. 2.7) as

$$f_{VI}(x) = e^x, \qquad 0 < x < 2\pi. \qquad (2.60)$$

The integrals are now a little more difficult to perform, but the method is the same. We shall first determine

$$C = \pi a_k = \int_0^{2\pi} e^x \cos kx \, dx = \frac{1}{k} e^x \sin kx \Big|_0^{2\pi} - \frac{1}{k} \int_0^{2\pi} e^x \sin kx \, dx$$

$$= -\frac{1}{k} \left\{ -\frac{1}{k} e^x \cos kx \Big|_0^{2\pi} + \frac{1}{k} \int_0^{2\pi} e^x \cos kx \, dx \right\} = \frac{e^{2\pi} - 1}{k^2} - \frac{C}{k^2}, \quad (2.61)$$

whence

$$C = \frac{e^{2\pi} - 1}{1 + k^2}. \qquad (2.62)$$

Similarly,

$$S = \pi b_k = \int_0^{2\pi} e^x \sin kx \, dx = -\frac{1}{k} e^x \cos kx \Big|_0^{2\pi} + \frac{1}{k} \int_0^{2\pi} e^x \cos kx \, dx$$

$$= -\frac{e^{2\pi} - 1}{k} + \frac{1}{k} \frac{e^{2\pi} - 1}{1 + k^2} = -\frac{k(e^{2\pi} - 1)}{1 + k^2}. \qquad (2.63)$$

The Fourier series is then obtained in the form

$$f_{VI}(x) = (e^{2\pi} - 1)\left[\frac{1}{2} + \sum_{k=1}^{\infty} \frac{1}{1 + k^2} (\cos kx - k \sin kx) \right]. \qquad (2.64)$$

2.4 How to sum up a series: a spin-off

I do not pretend for a moment that for an engineer this is an important part of the course on Fourier series. If an engineer needs to sum up a series, he/she will look it up in a handbook or will work it out on a computer. I feel, however, that it is worth mentioning this particular application of the Fourier series, mainly because it requires so little effort. It's a spin-off of the work already done.

Take, for example, the Fourier series of the function shown in Fig. 2.6 worked out in Example V of the previous section and given by eqn (2.59). At $x = 2\pi$ there is a discontinuity. The limit from the left-hand side is 1, whereas from the right-hand side it is zero. Hence the Fourier series should converge to 1/2 at $x = 2\pi$.

Substituting $x = 2\pi$ into eqn (2.59) we find that the sine terms vanish and all the cosine terms are equal to unity, leading to

$$\tfrac{1}{2} = \tfrac{1}{3} + \frac{1}{\pi^2} \sum_{k=1}^{\infty} \frac{1}{k^2}, \tag{2.65}$$

whence

$$S_1 = \sum_{k=1}^{\infty} \frac{1}{k^2} = \frac{\pi^2}{6}. \tag{2.66}$$

Neat, isn't it?

We can actually use the same Fourier series to find the sum of another series by putting $x = \pi$ into eqn (2.65). The value of the function is then 1/4, leading to

$$\tfrac{1}{4} = \tfrac{1}{3} + \frac{1}{\pi^2} \sum_{k=1}^{\infty} \frac{(-1)^k}{k^2}, \tag{2.67}$$

whence

$$S_2 = \sum_{k=1}^{\infty} \frac{(-1)^{k-1}}{k^2} = 1 - \tfrac{1}{4} + \tfrac{1}{9} - \tfrac{1}{16} + \cdots = \frac{\pi^2}{12} \tag{2.68}$$

If you like mathematical tricks, you will be interested to see that we can easily sum up a third series with the aid of eqns (2.66) and (2.68). We want to find the sum

$$S_3 = 1 + \tfrac{1}{9} + \tfrac{1}{25} + \cdots = \sum_{i=1}^{\infty} \frac{1}{(2i-1)^2}. \tag{2.69}$$

You may realize that

$$S_2 = 1 - \tfrac{1}{4} - \tfrac{1}{9} - \tfrac{1}{16} + \cdots = S_3 - \tfrac{1}{4}(1 + \tfrac{1}{4} + \tfrac{1}{9} + \cdots) = S_3 - \frac{S_1}{4}, \tag{2.70}$$

whence

$$S_3 = S_2 + \frac{S_1}{4} = \frac{\pi^2}{8}. \tag{2.71}$$

If you want to continue the fun you may attempt Exercises 2.6 and 2.7.

2.5   Differentiation and integration

If a function can be accurately represented by its Fourier series, then one may expect that a certain operation performed on the function can also be performed on its Fourier series. We actually assumed that much when we integrated the series term by term in order to find the coefficients.

Let us now do a few examples. First recall our functions $f_I(x)$ and $f_{II}(x)$ shown in Figs. 2.3 and 2.4 respectively. It may be seen by inspection that

$$\int_0^x f_I(x)\, dx = f_{II}(x)\,. \tag{2.72}$$

Is this true for their Fourier series? To find that out we can integrate eqn (2.42) term by term to yield

$$\int_0^x f_I(x)\, dx = \frac{4}{\pi} \sum_{i=1}^{\infty} \frac{1}{2i-1} \int_0^x \sin(2i-1)\, x\, dx$$

$$= -\frac{4}{\pi^2} \sum_{i=1}^{\infty} \frac{\cos(2i-1)x - 1}{(2i-1)^2} = -\frac{4}{\pi} \left[\sum_{i=1}^{\infty} \frac{\cos(2i-1)x}{(2i-1)^2} - \frac{\pi^2}{8} \right], \tag{2.73}$$

where we have used eqn (2.71). We may now recognize that eqn (2.73) is identical with eqn (2.46), i.e. the relationship given by eqn (2.72) is also valid for their Fourier series.

Will it work the other way round? Let's try it and differentiate eqn (2.46). We obtain

$$\frac{df_{II}}{dx} = \frac{4}{\pi} \sum_{i=1}^{\infty} \frac{\sin(2i-1)x}{2i-1}, \tag{2.74}$$

which may be seen to be identical with eqn (2.42).

Will it always work that nicely? As our next example, let us integrate eqn (2.47) between the limits 0 and x. It yields

$$\int_0^x \frac{x}{2\pi}\, dx = \frac{x^2}{\pi}\,. \tag{2.75}$$

Thus, integrating the Fourier series given by eqn (2.51) and dividing it by π should yield the Fourier series given by eqn (2.59).

The integration of the right-hand side of eqn (2.51) yields

$$\int_0^x f_{III}(x)\, dx = \frac{x}{2} + \frac{1}{\pi} \sum_{k=1}^{\infty} \frac{\cos kx - 1}{k^2}\,. \tag{2.76}$$

But from the definition of $f_{II}(x)$ (eqn (2.46)),

$$x = 2\pi \left[\frac{1}{2} - \frac{1}{\pi} \sum_{k=1}^{\infty} \frac{1}{k^2} \sin kx \right], \tag{2.77}$$

and from Eqn (2.66)

$$\sum_{k=1}^{\infty} \frac{1}{k^2} = \frac{\pi^2}{6}. \tag{2.78}$$

Substituting eqns (2.77) and (2.78) into eqn (2.76) we find

$$\int_0^x f_{\text{III}}(x)\,\mathrm{d}x = \frac{\pi}{2} + \frac{1}{\pi}\sum_{k=1}^{\infty}\left(\frac{\cos kx}{k^2} - \frac{\pi}{k}\sin kx\right) - \frac{\pi}{6}, \tag{2.79}$$

which, divided by π, will indeed yield the Fourier series of eqn (2.59).

Will this work the other way round? Let's try it and differentiate eqn (2.59). We obtain

$$\frac{\mathrm{d}f_V(x)}{\mathrm{d}x} = \frac{1}{\pi^2}\sum_{k=1}^{\infty}\left(-\frac{\sin kx}{k} - \pi\cos x\right). \tag{2.80}$$

The first term on the right-hand side is indeed part of the required result, but we have in addition

$$-\frac{1}{\pi}\sum_{k=1}^{\infty}\cos kx. \tag{2.81}$$

What can we say about the above sum? We can immediately say that the series is not convergent. Why not? Because the kth term does not tend to zero as k tends to infinity.

Is something wrong with the Fourier series of x^2? In order to see, we plot the Fourier series for the first 20 components (denoted by f_{20} following the notation introduced in Section 1.5) in Fig. 2.8. It does not seem worse than the others we have seen before, but concerning differentiability it looks bad enough. There are fast-varying ripples of not-negligible amplitudes at the two edges.

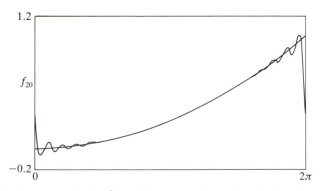

Fig. 2.8 The function $(x/2\pi)^2$ and its approximation by its Fourier series containing the constant term plus the first 20 harmonics

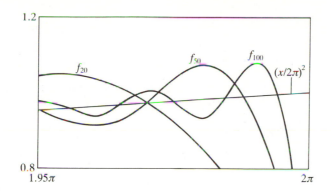

Fig. 2.9 The function $(x/2\pi)^2$ and its approximations in the interval 1.95π to 2π

Let us just look at one of the edges in more detail. In Fig. 2.9 we plot the Fourier series for $N = 20$, 50 and 100 in the interval $1.95\pi < x < 2\pi$. As N increases, the ripples become steeper but their amplitudes do not decrease much. In fact the amplitude of the last ripple (this is the ubiquitous Gibbs phenomenon again; for an analysis see the Appendix) stays the same. It is easy to see then that the differential will vary between large negative and large positive values as N tends to infinity and, in particular, the differential must tend to infinity at $x = 2\pi$.

So what is our conclusion? We just need to use common sense. $f_{\mathrm{I}}(x)$ of Fig. 2.3 can be integrated all right and $f_{\mathrm{II}}(x)$ of Fig. 2.4 can be differentiated all right, that much is obvious. Trying to differentiate $f_{\mathrm{V}}(x)$ of Fig. 2.5b was, however, a little ambitious. $f_{\mathrm{V}}(x)$ is not differentiable. It was too much to expect that its Fourier series could be differentiated. So we could simply say that whether a Fourier series can be differentiated or not depends on whether the original periodic function is differentiable or not. Integration of the series, on the other hand, is always permissible.

2.6 The coefficients for arbitrary period

Up to now we have always considered a period of 2π. Naturally, whatever we have said so far will also apply to an arbitrary period; we only need to make a few minor adjustments in the formulae.

I could carry on using the x coordinate, but out of force of habit I normally use t as the variable when it comes to arbitrary period. Denoting the period by T, the fundamental components are $\cos(2\pi t/T)$ and $\sin(2\pi t/T)$. This is because at $t = T$ the argument is 2π. The Fourier series may then be written in the form

$$f(t) = \frac{a_0}{2} + \sum_{k=1}^{\infty} a_k \cos \frac{2\pi kt}{T} + \sum_{k=1}^{\infty} b_k \sin \frac{2\pi kt}{T}. \tag{2.82}$$

Let us follow again the same routine as in Section 4.1: multiply eqn (2.82) by $\cos(2\pi lt/T)$ and integrate for one period. We obtain

$$\int_0^T f(t) \cos \frac{2\pi lt}{T} \, dt = \frac{a_0}{2} \int_0^T \cos \frac{2\pi lt}{T} \, dt + \sum_{k=1}^{\infty} a_k \int_0^T \cos \frac{2\pi kt}{T} \cos \frac{2\pi lt}{T} \, dt$$

$$+ \sum_{k=1}^{\infty} b_k \int_0^T \sin \frac{2\pi kt}{T} \cos \frac{2\pi lt}{T} \, dt. \tag{2.83}$$

The relevant orthogonality relations are now

$$\int_0^T \cos \frac{2\pi kt}{T} \cos \frac{2\pi lt}{T} \, dt = \begin{matrix} 0 \\ T/2 \end{matrix} \quad \text{if} \quad \begin{matrix} l \neq k \\ l = k \end{matrix} \tag{2.84}$$

and

$$\int_0^T \sin \frac{2\pi kt}{T} \cos \frac{2\pi lt}{T} \, dt = 0. \tag{2.85}$$

Using the above relations we obtain from eqn (2.83)

$$a_k = \frac{2}{T} \int_0^T f(t) \cos \frac{2\pi kt}{T} \, dt, \tag{2.86}$$

a formula which applies for $k = 0$ as well.

By similar technique we obtain for the coefficient of the sine terms

$$b_k = \frac{2}{T} \int_0^T f(t) \sin \frac{2\pi kt}{T} \, dt. \tag{2.87}$$

Equations (2.86) and (2.87) above are generalizations of eqns (2.18) and (2.21) for arbitrary period. The symmetry properties can of course still be taken into account, to result in simplified formulae.

To illustrate the derivation of the coefficients for arbitrary period I shall now give two examples, both of them taken from engineering practice.

We may describe the variation of voltage impressed upon a circuit by a voltage generator as

$$V = V_0 \cos \omega t, \qquad \omega = 2\pi/T. \tag{2.88}$$

In our first example we wish to find the Fourier series of the function obtained from eqn (2.88) by half-wave rectification. This means that the negative part of the cosine function is chopped off, i.e. we wish to find

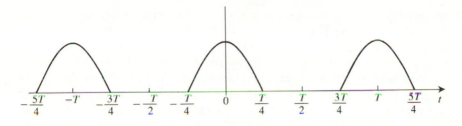

Fig. 2.10 A half-wave rectified cosine function

the Fourier expansion of the periodic function shown in Fig. 2.10. This is an even function; there will therefore be only cosine components in the expansion. The coefficients can be determined from the formula

$$a_k = \frac{4}{T} \int_0^{T/2} f(t) \cos \frac{2\pi k t}{T} \, dt, \qquad (2.89)$$

leading (this is now left as an exercise) to the Fourier series

$$f(t) = \frac{1}{\pi} + \frac{1}{2} \cos \frac{2\pi t}{T} - \frac{2}{\pi} \sum_{i=1}^{\infty} \frac{(-1)^i}{4i^2 - 1} \cos \frac{4\pi i t}{T}. \qquad (2.90)$$

As our second example we take the function obtained from the cosinusoidal by full-wave rectification. The curve is obtained (Fig. 2.11a)

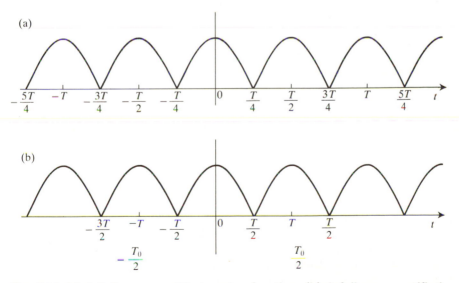

Fig. 2.11 (a) A full-wave rectified cosine function. (b) A full-wave rectified cosine function with different notations

by turning the negative parts into positive ones: mathematically speaking, we wish to find the Fourier series of the function $|\cos(2\pi t/T)|$. We have to be very careful now. I know from experience that many undergraduates get utterly confused when trying to find the coefficients. The source of confusion is that T in the expression $\cos(2\pi t/T)$ is the period of the original (before rectification) function $\cos(2\pi t/T)$, whereas the period of the periodic function (Fig. 2.11a) for which we want to find the Fourier series is $T/2$.

The confusion can be cleared by changing our notations. If we want to use eqns (2.86) or (2.87), we should realize that T means the period of the periodic function. So in trying to find the Fourier series the notation used in Fig. 2.11a is misleading. We should change the notation for the period of our original cosinusoidal to T_0 as shown in Fig. 2.11b, and the period of the periodic function is then $T = T_0/2$.

The situation should now be clear. We wish to find the Fourier series of the function

$$f(t) = \cos\frac{\pi t}{T} \tag{2.91}$$

and in order to find the coefficients we need to integrate for one period of the periodic function, that is from $-T/2$ to $T/2$. However, eqn (2.91) being even, it is sufficient to integrate for a half-period, for 0 to $T/2$. Thus the coefficients are given by

$$a_k = \frac{4}{T}\int_0^{T/2} \cos\frac{\pi t}{T} \cos\frac{2\pi kt}{T}\, dt. \tag{2.92}$$

Investing a little mathematical labour we end up with the Fourier series

$$f(t) = \frac{2}{\pi} - \frac{4}{\pi}\sum_{k=1}^{\infty} \frac{(-1)^k}{4k^2-1} \cos\frac{2\pi kt}{T} \tag{2.93}$$

(and remember, $T = T_0/2$).

As our third example we shall take a rather curious function, a square wave modulation of a high-frequency signal. In mathematical terms our function is defined as follows

$$f(t) = \begin{matrix} \sin \omega_0 t \\ 0 \end{matrix} \quad \text{for} \quad \begin{matrix} 0 < t < T/2 \\ T/2 < t < T, \end{matrix} \tag{2.94}$$

and then repeated with the same period T as shown schematically in Fig. 2.12. In practice ω_0 could be very high. Since at some stage I wish to plot this function, I shall choose it only relatively high, namely 100 times higher than the fundamental frequency, i.e.

$$\omega_0 = 100\omega = 100\frac{2\pi}{T}. \tag{2.95}$$

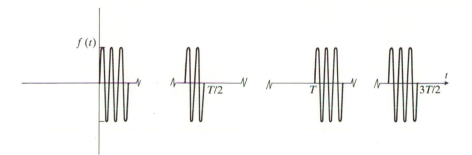

$f(t)$

$T/2$ T $3T/2$ t

Fig. 2.12 A square wave modulated sinusoidal defined by eqn (2.94)

So we have exactly fifty periods of a sine wave from $t = 0$ to $t = T/2$ and then nothing for the same length of time.

What kind of Fourier components should we expect? In all the examples considered so far, the fundamental component always made the greatest contribution, and the contribution of the harmonics declines (not necessarily monotonically) with increasing harmonic number. Will it be the same now? Common sense would say, no. It is hardly conceivable that the fundamental component could much contribute to approximating our function. By common sense the harmonics that matter should be close to the 100th harmonic, which is the frequency of the sinusoidal in the interval 0 to $T/2$.

We shall now see whether our expectations are right. By applying the rules we obtain the following Fourier series:

$$ f(t) = -\frac{2}{\pi} \sum_{i=1}^{\infty} \frac{2i - 1}{100^2 - (2i - 1)^2} \cos \frac{2\pi(2i - 1)t}{T} + \tfrac{1}{2} \sin \frac{200\pi t}{T} \quad (2.96) $$

It is certainly not the run-of-the-mill Fourier series. The cosine series may be seen to give large contribution only in the region where $2i - 1$ is close to 100, and the sine series has only one single component at the 100th harmonic.

Let us again plot the Fourier series for the first N harmonics. It would need quite a lot of paper to plot the function for the whole period 0 to T. The crucial part of this function, however, is in the vicinity of $t = 0.5T$. Hence we shall look at the range $0.475T < t < 0.525T$. Between $t = 0.475T$ and $t = 0.5T$ there should be two-and-a-half periods of the sinusoidal and the function should be zero beyond $t = 0.5T$. The approximation to $f(t)$ for $N = 90$ (denoted by f_{90}) is shown in Fig. 2.13a. There is absolutely no indication as yet of radical changes at $t = 0.5T$. However for $N = 110$ (Fig. 2.13b) there is clear indication of the

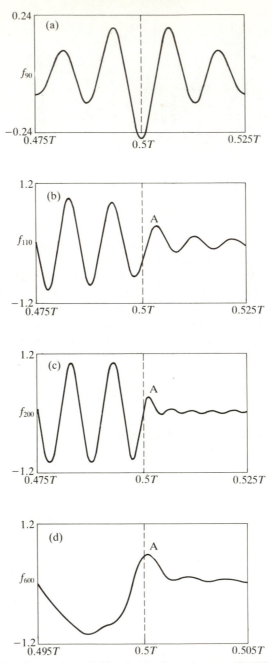

Fig. 2.13 The Fourier series of Fig. 2.12 given by eqn (2.96). Parts (a), (b) and (c) show the sum of the first 90, 110 and 200 harmonics, respectively, plotted in the interval $0.475T$ to $0.525T$. (d) The sum of the first 600 harmonics plotted in the interval $0.495T$ to $0.505T$

two-and-a-half periods of the sinusoidal and the subsequent decay beyond $t = 0.5T$, but note the overshoot at A. For $N = 200$ the approximate function f_{200} is plotted in Fig. 2.13c. The main features are the same as before but the ripples beyond $t = 0.5T$ are now smaller and the position of the overshoot has moved closer to $t = 0.5T$. If we take as many as $N = 600$ components then the decay is even faster (see Fig. 2.13d and note that the horizontal scale has been expanded by a factor of 5) but the overshoot is still there. It is the Gibbs phenomenon all over again, we cannot get rid of it.

2.7 The Fourier series as optimum approximation

As our examples have shown, we have been doing pretty well with our (or rather Fourier's) series. But is it the best we can do? Could we not obtain a better approximation by choosing some other set of coefficients? How should we determine those coefficients? And, anyway, how should we measure how good the approximation is?

The measure we shall use is the average measure discussed in Section 1.2. If $f(t)$ is the function to be approximated and

$$f_N(t) = \frac{a_0}{2} + \sum_{k=1}^{N} a_k \cos \frac{2\pi kt}{T} + \sum_{k=1}^{N} b_k \sin \frac{2\pi kt}{T} \tag{2.97}$$

is a series containing N components, then the expression to be minimized is

$$\int_0^T [f(t) - f_N(t)]^2 \, dt = \text{Min.} \tag{2.98}$$

Minimized as a function of what? The variables are now the coefficients a_k and b_k. The procedure to follow is to differentiate partially eqn (2.98) by a_k and b_k for each value of k and equate the result with zero. Let's do it formally:

$$\frac{\partial}{\partial a_k} \int_0^T [f(t) - f_N(t)]^2 \, dt = 2 \int_0^T [f(t) - f_N(t)] \frac{\partial f_N(t)}{\partial a_k} \, dt$$

$$= 2 \int_0^T [f(t) - f_N(t)] \cos \frac{2\pi kt}{T} \, dt. \tag{2.99}$$

Applying now the orthogonality relationships, the above equation reduces to

$$2 \int_0^\pi \left[f(t) - a_k \cos \frac{2\pi k}{T} t \right] \cos \frac{2\pi kt}{T} \, dt, \tag{2.100}$$

which is further equal to

$$2\left\{\int_0^T f(t) \cos\frac{2\pi kt}{T}\, \mathrm{d}t - a_k\frac{T}{2}\right\}, \tag{2.101}$$

which is equal to zero when

$$a_k = \frac{2}{T}\int_0^T f(t) \cos\frac{2\pi kt}{T}\, \mathrm{d}t. \tag{2.102}$$

Thus we have got the result that if we set out to solve the optimization problem presented in eqn (2.98) the coefficients we get are the very same as those in the Fourier series. The same applies to a_0 and b_k as well, yielding the remarkable result that the Fourier series is the best approximation in the sense that it minimizes the error-squared averaged over a period.

2.8   Full-range, half-range, quarter-range, etc.

We have so far talked about approximating a periodic function with the aid of sine and cosine functions, and this is indeed what the Fourier series is all about. But is it really necessary for the function to be periodic? Take for example the function shown in Fig. 2.14a, defined as

$$f(t) = -t^2 + 2t. \tag{2.103}$$

The main point in the present case is that it is of no interest at all what values this function will take outside the 0,1 interval. The physical problem is such that t can be only between 0 and 1. So $f(t)$ is definitely not a periodic function but we could still pretend that it is periodic, as shown in Fig. 2.14b, and then just turn the handle and determine the coefficients from eqns (2.89) and (2.90).

But if the function is of interest only between 0 and 1 then we can continue it outside this range in any way we fancy. Thus a possible choice is to regard $f(t)$ as an odd function, continue it accordingly in the interval -1 to 0, and then regard the range -1 to 1 as the period, and make a periodic function out of it as shown in Fig. 2.14c.

Once we strike on the idea of choosing an odd function, it is easy to think of continuation in the form of an even function. So the periodic function shown in Fig. 2.14d is another logical choice.

Summarizing what we have done so far, we may call our choice in Fig. 2.14b full-range (since the originally specified range is the same as the period of the chosen periodic function), and our choices in Fig. 2.14c and d half-range (since the originally specified range is half of the period of the chosen periodic function).

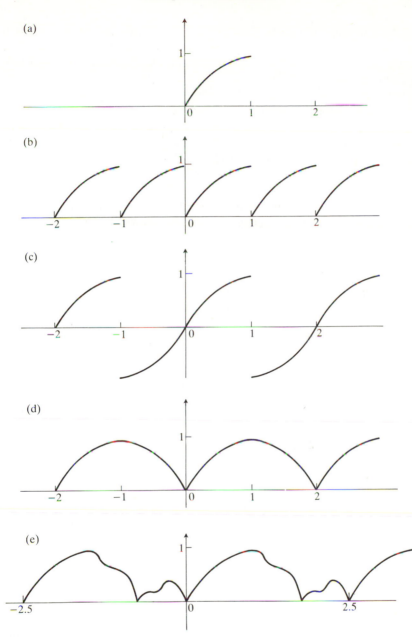

Fig. 2.14 (a) The function $f(t) = -t^2 + 2t$ defined in the interval 0 to 1. (b) Full-range periodic extension of the function shown in (a). (c) Half-range, odd periodic extension of the function shown in (a). (d) Half-range even periodic extension of the function shown in (a). (e) An arbitrary periodic extension of the function shown in (a)

I wish to emphasize once more that we are entirely free to choose our periodic function outside the range 0 to 1. Thus a possible choice is the function shown in Fig. 2.14e, although I'm not sure at the moment that it is good for anything.

What should dictate our choice? The rate of convergence could be one consideration; the suitability of the resulting function for further operations could be another consideration. The latter problem will be amply illustrated in Chapter 4, where the solution of a number of partial differential equations will be discussed. The rate of convergence, the way the specified function is being built up for various continuations, will be discussed in the present section.

So let us see first the Fourier series for the full-range case. The coefficients may be determined from the formulae

$$a_k = 2 \int_0^1 (-t^2 + 2t) \cos 2\pi kt \, dt \quad \text{and} \quad b_k = 2 \int_0^1 (-t^2 + 2t) \sin 2\pi kt \, dt,$$

(2.104)

yielding the series

$$f(t) = \frac{2}{3} - \frac{1}{\pi^2} \sum_{k=1}^{\infty} \left[\frac{1}{k^2} \cos 2\pi kt + \frac{\pi}{k} \sin 2\pi kt \right].$$
(2.105)

How fast does the series converge? The results for $N = 1$, 3, 5 and 10 components are shown in Fig. 2.15a, b, c, and d respectively. the approximation looks quite good for five components and even better for 10 components, but there are difficulties at the edges. We have here the Gibbs phenomenon once more, since there are discontinuities at $t = 0$ and $t = 1$ and the Fourier series converges at both places to $(0 + 1)/2 = 0.5$.

For the odd half-range series there are only sine components. The coefficients are calculated from

$$b_k = 2 \int_0^1 (-t^2 + 2t) \sin \pi kt \, dt,$$
(2.106)

leading to the Fourier series

$$f(t) = \sum_{k=1}^{\infty} \left\{ -\frac{2}{\pi k} (-1)^k + \frac{4}{\pi^3 k^3} [1 - (-1)^k] \right\} \sin \pi kt.$$
(2.107)

Do we have faster convergence in the 0,1 interval with this series? The approximations for $N = 1$, 3, 5 and 10 are shown in Fig. 2.16a, b, c, and d respectively. Well, at $t = 0$ the situation has improved because the function is now continuous there, but there is considerable deterioration around $t = 1$ since the discontinuity is now larger there. On the whole we might say (if our measure is the maximum error for a given number of

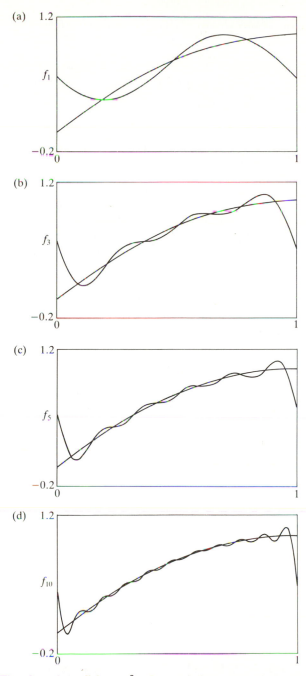

Fig. 2.15 The function $f(t) = -t^2 + 2t$, and the sum of its first N Fourier harmonics + constant term for the full-range periodic extension of Fig. 2.14b for (a) $N = 1$. (b) $N = 3$, (c) $N = 5$, and (d) $N = 10$

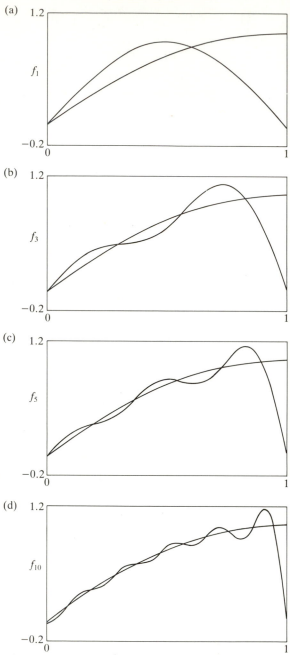

Fig. 2.16 The function $f(t) = -t^2 + 2t$, and the sum of its first N Fourier harmonics for the half-range odd periodic extension of Fig. 2.14c for (a) $N = 1$, (b) $N = 3$, (c) $N = 5$, and (d) $N = 10$

components) that having a larger discontinuity we are worse off. A possible reason for preferring the odd half-range function stems from a quite different argument. We may choose it because the Fourier series then converges to zero at $t = 0$, and in some problems that may be the physical requirement.

Next, we shall look at the even half-range function. There are then only cosine coefficients, calculated from

$$a_k = \int_0^1 (-t^2 + 2t) \cos \pi kt \, dt, \qquad (2.108)$$

which yield the Fourier series

$$f(t) = \frac{2}{3} - \frac{4}{\pi^2} \sum_{k=1}^{\infty} \frac{1}{k^2} \cos \pi kt. \qquad (2.109)$$

The convergence may now be expected to be much better because the even function (Fig. 2.14d) is continuous. The approximation with the Fourier series is now shown only for $N = 1$, 2 and 3 (Fig. 2.17a, b, and c) because beyond that the approximation becomes better than the thickness of the line. It is worth noting that the amplitude of the kth component declines as $1/k^2$, in contrast to the previous two cases when they declined (or at least one of the terms declined) as $1/k$. So there is no doubt that for the even function we have better convergence. The point where the approximation is not so good is $t = 0$, because the derivative of the function has a discontinuity there.

Having discussed full-range and half-range, should we look at some other continuations as well? We may ask, for example, what continuation is likely to give even faster convergence? Well, the best recipe seems to be to keep both the function and the derivative continuous at each point. We could achieve that by choosing our period to be between $t = -2$ and 2 (see Fig. 2.18) and define the function as

$$f(t) = \begin{matrix} t^2 + 2t \\ -t^2 + 2t \end{matrix} \quad \text{if} \quad \begin{matrix} -2 < t < 0 \\ 0 < t < 2. \end{matrix} \qquad (2.110)$$

This is now an odd function and, in fact, very much resembles a sinusoidal. One may expect therefore that the fundamental component will already give a good approximation.

Let's do the calculation. The symmetry properties of the above function correspond to one of those discussed in Section 2.2, namely when the function is odd, and its part for positive coordinates is even about the middle. The result is that there are only odd sinusoidal components present. The coefficients can be calculated from

$$b_{2i-1} = \frac{8}{T} \int_0^1 (-t^2 + 2t) \sin \left[\frac{\pi}{2} (2i - 1)t \right] dt, \qquad (2.111)$$

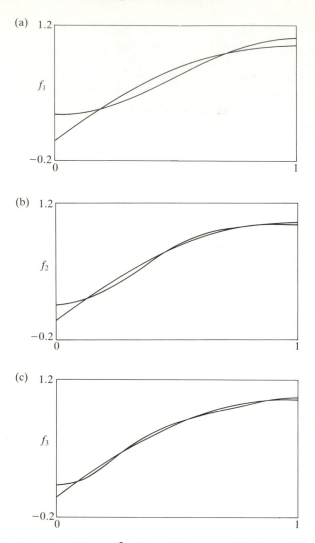

Fig. 2.17 The function $f(t) = -t^2 + 2t$ and the sum of its first N harmonics + constant term for the half-range even periodic extension of Fig. 2.14d for (a) $N = 1$, (b) $N = 2$, and (c) $N = 3$

yielding the Fourier series

$$f(t) = \frac{32}{\pi^3} \sum_{i=1}^{\infty} \frac{1}{(2i-1)^3} \sin \left[\frac{\pi}{2}(2i-1)t \right] \tag{2.112}$$

We find that the fundamental component will indeed give a very good approximation, as shown in Fig. 2.19. The next term ($i = 3$) has an

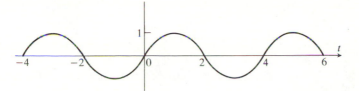

Fig. 2.18 Quarter-range, odd periodic extension of the function shown in Fig. 2.14a

amplitude of only about 4% of the fundamental component. Adding that term, the approximation becomes so good that it is within the thickness of the line in the whole range. The convergence is really fast. It goes now as $1/k^3$.

Should we consider any further continuation? There is one type I would still like to consider, because it will provide a convenient introduction to the so-called Fourier integral, something I would like to mention but do not wish to dwell on to any length as it is beyond the scope of the present lectures.

We shall now choose what may be called the simplest continuation, namely we choose the function to be zero outside the interval 0,1. The only remaining problem is then to choose the period. An example when $T = 10$ is shown in Fig. 2.20.

We shall now do the derivation for a general T, i.e. the function is defined as

$$f(t) = \begin{array}{cc} -t^2 + 2t \\ 0 \end{array} \quad \text{if} \quad \begin{array}{c} 0 < t < 1 \\ 1 < t < T, \end{array} \quad (2.113)$$

and is continued periodically outside the region 0,T. The coefficients of

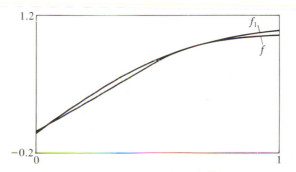

Fig. 2.19 The function $f(t) = -t^2 + 2t$ and the fundamental component of the periodic function shown in Fig. 2.18

Fig. 2.20 Full-range periodic extension of the function defined by eqn (2.113) taking $T = 10$

the Fourier series may now be determined from the formulae

$$a_k = \frac{2}{T} \int_0^1 (-t^2 + 2t) \cos \frac{2\pi kt}{T} \, dt, \qquad b_k = \frac{2}{T} \int_0^1 (-t^2 + 2t) \sin \frac{2\pi kt}{T} \, dt,$$

$$(2.114)$$

leading to the Fourier series

$$f(t) = \frac{2}{3} + \frac{2}{T} \sum_{k=1}^{\infty} [A(k\omega) \cos k\omega t + B(k\omega) \sin k\omega t] \qquad (2.115)$$

where

$$A(k\omega) = -\frac{2}{k^2\omega^2} + \left[1 + \frac{2}{(k\omega)^2}\right] \frac{\sin k\omega}{k\omega},$$

$$(2.116)$$

$$B(k\omega) = \frac{2}{(k\omega)^3} - \left[1 + \frac{2}{(k\omega)^2}\right] \frac{\cos k\omega}{k\omega}.$$

What can we expect now concerning convergence? The function varies from 0 to 1 and the period is from 0 to T, i.e. the period is T times larger than the interval in which the function is non-zero. We may now use a simple argument suggesting that in order to approximate the variation of the function in the interval $(0,1)$ the highest Fourier component should have a period roughly coinciding with that interval, i.e. we may persuade ourselves that $N = T$ will yield a reasonable approximation.

For $T = 10$ the Fourier series is plotted for $N = 5$, 10 and 20 in Fig. 2.21. As may be seen, there is little resemblance to the original function for $N = 5$ but, as expected, $N = 10$ gives a reasonable approximation which improves further for $N = 20$.

Very similar conclusions may be drawn from Fig. 2.22 for $T = 100$. Then $N = 50$ is inadequate: $N = 100$ gives good resemblance which is further improved for $N = 200$.

Up to now we have always asked the question: how good is the approximation for a given number of terms? Let me reformulate the question in a slightly different form. If we have a signal for a duration of τ ($\tau = 1$ is chosen in the present example) and period T, what is the maximum frequency we need to include in order to have a reasonable

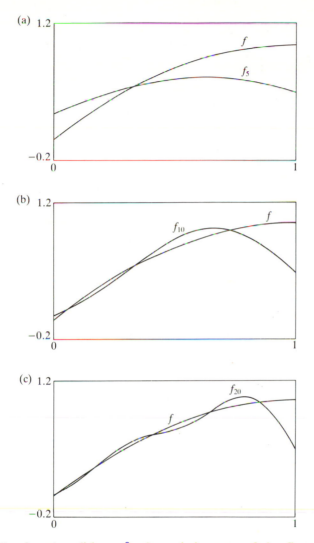

Fig. 2.21 The function $f(t) = -t^2 + 2t$ and the sum of the first N Fourier harmonics + constant term of the periodic function shown in Fig. 2.20 for (a) $N = 5$, (b) $N = 10$, and (c) $N = 20$

chance to recognize the signal. Notice the difference: I am not asking how many terms that will involve. I am interested only in the maximum frequency.

Let us now introduce the amplitude of the kth component, d, as the quadrature of the sine and cosine components, i.e.

$$d_k = (a_k^2 + b_k^2)^{1/2}, \qquad (2.117)$$

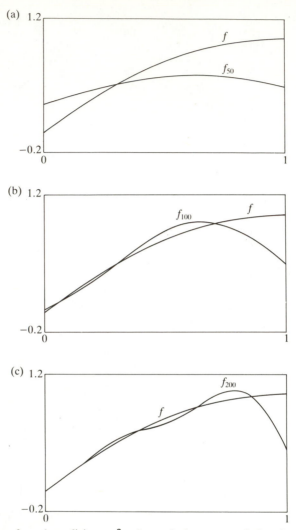

Fig. 2.22 The function $f(t) = -t^2 + 2t$ and the sum of the first N Fourier harmonics + constant term of the function defined by eqn (2.113) taking $T = 100$ for (a) $N = 50$, (b) $N = 100$, and (c) $N = 200$

and plot $d_k T/2$ in Figs. 2.23a and b against k for $T = 10$ and 100 respectively. In Fig. 2.23a the first 30 components are plotted, whereas in Fig. 2.23b we plot the components up to 300. Looking at these two figures it is clear now that the envelope is the same in both cases. For example, $k = 20$ in Fig. 2.23a yields the same amplitude as $k = 200$ in Fig. 2.23b, and the corresponding frequency is $20 \times (2\pi/10) = 4\pi$ in the first case, and exactly the same, $200 \times (2\pi/100) = 4\pi$ in the second case.

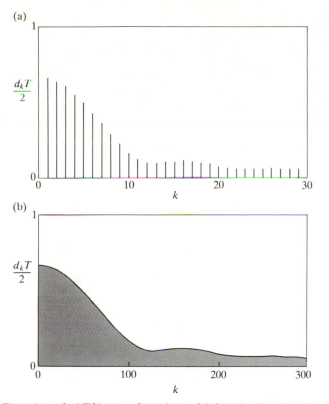

Fig. 2.23 The plot of $d_k T/2$ as a function of k for the Fourier harmonics of the function defined by eqn (2.113) for (a) $T = 10$ and (b) $T = 100$

So it seems reasonable to generalize from what we have seen so far and say that, as T increases, we need to include more and more components in order to have the same degree of approximation but the maximum frequency is always the same.

What happens as T tends to infinity? The most important conclusion we can come to is that the frequency spectrum tends towards a continuous one. As T tends to infinity the amplitudes of the individual components, d_k, must of course tend to zero, but $d_k T/2$, remarkably, remains finite, so that the envelope may be defined as a new function, and this is known as the Fourier transform of the function, which in the present case was defined in the $(0, 1)$ interval and (since we permitted T to tend to infinity) is not periodic.

2.9 Exponential form

There is a tendency in electrical engineering (and in some other branches of engineering concerned with vibrations) to use complex exponential forms instead of trigonometric ones. The reason is that the differential equations governing the behaviour of a large number of devices are linear, in which case the exponential notation leads to much simpler mathematics and also to a more intuitive concept of phase angle.

It turns out that the Fourier series can also be put into a more compact form by using exponential notation. For this purpose we shall use the relationships

$$\cos\frac{2\pi kt}{T} = \frac{1}{2}\left[\exp\left(j\frac{2\pi kt}{T}\right) + \exp\left(-j\frac{2\pi kt}{T}\right)\right] \qquad (2.118)$$

and

$$\sin\frac{2\pi kt}{T} = \frac{1}{2j}\left[\exp\left(j\frac{2\pi kt}{T}\right) - \exp\left(-j\frac{2\pi kt}{T}\right)\right], \qquad (2.119)$$

which may then be substituted into eqn (2.82) to yield

$$f(t) = \frac{1}{2}\left\{a_0 + \sum_{k=1}^{\infty}\left[(a_j - jb_k)\exp\left(j\frac{2\pi kt}{T}\right) + (a_k + jb_k)\exp\left(-j\frac{2\pi kt}{T}\right)\right]\right\} \qquad (2.120)$$

Introducing the further notations

$$c_0 = \frac{a_0}{2}, \qquad c_k = \tfrac{1}{2}(a_k - jb_k), \qquad c_{-k} = \tfrac{1}{2}(a_k + jb_k) = c_k^*, \quad (2.121)$$

eqn (2.120) may be written in the compact form

$$f(t) = \sum_{k=-\infty}^{\infty} c_k \exp\left(j\frac{2\pi kt}{T}\right). \qquad (2.122)$$

We shall now, once more, derive a formula for the coefficients. We multiply both sides of eqn (2.122) by $\exp(-j2\pi lt/T)$ and integrate for one period. We obtain

$$\int_0^T f(t)\exp\left(-j\frac{2\pi lt}{T}\right)dt = \sum_{k=-\infty}^{\infty}\int_0^T \exp\left[j\frac{2\pi t}{T}(k-l)\right]dt. \quad (2.123)$$

The orthogonality relationship is now

$$\int_0^T \exp\left[j\frac{2\pi t}{T}(k-l)\right]dt = \begin{matrix} 0 \\ T \end{matrix} \quad \text{if} \quad \begin{matrix} k \neq l \\ k = l \end{matrix} \qquad (2.124)$$

leading to the following form for the coefficients

$$c_k = \frac{1}{T}\int_0^T f(t)\exp\left(-j\frac{2\pi kt}{T}\right)dt. \qquad (2.125)$$

It may be easily seen that the "old" coefficients are obtained from c_k

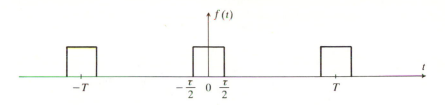

Fig. 2.24 A sequence of rectangular pulses of duration τ

using the relations

$$a_k = 2 \operatorname{Re} c_k \quad \text{and} \quad b_k = -2 \operatorname{Im} c_k. \quad (2.126)$$

Let us now see an example of how one could find the Fourier series in the new notation. The periodic function chosen is a sequence of rectangular pulses as shown in Fig. 2.24. We have already come across special cases of this function in Fig. 1.6 and in Fig. 2.3. The present function is more general in the sense that both τ, the duration of the pulse, and T, the period of the function, are arbitrary.

The coefficients may be obtained from eqn (2.125), noting that the period in this instance is taken from $-T/2$ to $T/2$. We obtain

$$c_k = \frac{1}{T} \int_{-\tau/2}^{\tau/2} \left(-j \frac{2\pi kt}{T} \right) dt = \frac{1}{T} \frac{1}{-j(2\pi k/T)} \left[\exp\left(-j \frac{\pi k\tau}{T} \right) - \exp\left(j \frac{\pi k\tau}{T} \right) \right]$$

$$= (1/\pi k) \sin(\pi k\tau/T) \quad (2.127)$$

What can we expect for convergence? How many components should we include to have a reasonable approximation? Well, we could use the argument advanced in the previous section and take $N = T/\tau$. Another often-used argument may be based on the construction shown in Fig. 2.25. If we want to approximate the pulse shape of duration τ then we need at least a sinusoidal of period 2τ. The period of the kth component is T/k, whence the highest component we need is $k = T/2\tau$.

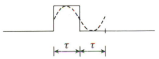

Fig. 2.25 Illustration of the simple argument that in order to be able to approximate a pulse of duration τ we need at least a sinusoidal of period 2τ

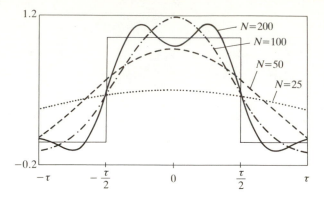

Fig. 2.26 The function of Fig. 2.24 plotted in the interval $-\tau$ to τ, and the sum of the first N Fourier harmonics + constant term taking $T/\tau = 100$ for $N = 25, 50, 100,$ and 200

Let's now do the calculations. Choosing $T/\tau = 100$ the original function and the Fourier series for $N = 25, 50, 100$ and 200 are plotted in Fig. 2.26 for the interval $(-\tau, \tau)$. For $N = 25$ one could hardly say that something specific is happening in the $(-\tau/2, \tau/2)$ interval. For $N = 50$ there is still no resemblance to the square shape, but at least we can see that the series has considerably higher values within the required interval than outside it. For $N = 100$ the distinction is even sharper but it is only for $N = 200$ that we can claim that there is some resemblance to the square shape.

Let us next look at the coefficients. We plot $100\,|d_k|$ in Fig. 2.27 as a

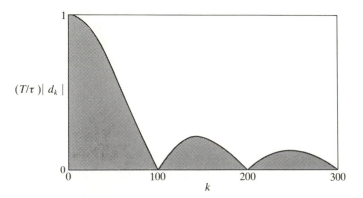

Fig. 2.27 Plot of the normalized Fourier coefficients $100\,|d_k|$ for the function of Fig. 2.24 as a function of k for $T/\tau = 100$

function of k. The first zero may be seen to occur at $k = T/\tau = 100$. It is very tempting to say that if we take into account all the components up to the first zero of the function then we have a reasonable approximation.

It needs to be stressed, however, that reasonableness is in the eye of the beholder. Several criteria may be used and it is fairly subjective which one of them is acceptable. In electrical engineering the criterion comes about as a consequence of a string of compromises based on technical, economic, and often political considerations.

2.10 Parseval's theorem

This is a relationship of great value to applied scientists because it relates quadratic forms to each other. In Sections 3.5 and 3.6 we shall show in specific examples how these quadratic forms can be represented as power. For the time being let us just do the mathematics. We shall write the Fourier series in the exponential form of eqn (2.122):

$$f(t) = \sum_{k=-\infty}^{\infty} c_k \exp\left(j\frac{2\pi kt}{T}\right). \tag{2.128}$$

Multiply both sides by $2f(t)/T$ and integrate them for the period 0 to T. We obtain

$$\frac{2}{T}\int_0^T [f(t)]^2 \, dt = \frac{2}{T} \sum_{k=-\infty}^{\infty} c_k \int_0^T f(t) \exp\left(j\frac{2\pi kt}{T}\right) dt \tag{2.129}$$

However, it may be seen from eqn (2.125) that the term under the integration sign on the right-hand side is equal to $c_k^* T$, with which we obtain the form (note that $|c_k|^2 = |c_{-k}|^2$)

$$\frac{2}{T}\int_0^T [f(t)]^2 \, dt = 2 \sum_{k=-\infty}^{\infty} |c_k|^2 = 2\left[c_0^2 + 2\sum_{k=1}^{\infty} |c_k|^2 \right]$$

$$= \frac{a_0^2}{2} + \sum_{k=1}^{\infty} (a_k^2 + b_k^2), \tag{2.130}$$

which is known as Parseval's theorem.

What can we use Parseval's theorem for? Apart from its representation of power in physical systems, one obvious use of it is to sum up series. You will remember that we managed to sum the series

$$\sum_{i=1}^{\infty} \frac{1}{(2i-1)^2} \tag{2.131}$$

in Section 2.4 by employing a mathematical trick. Now we can do it without any trick at all.

The expansion of the square wave shown in Fig. 2.3 is given by eqn (2.42). There are only odd sine terms with coefficients

$$b_{2i-1} = \frac{4}{\pi} \frac{1}{2i-1}. \tag{2.132}$$

Applying Parseval's theorem in the form of eqn (2.130) we find

$$2 = \frac{16}{\pi^2} \sum_{i=1}^{\infty} \frac{1}{(2i-1)^2}, \tag{2.133}$$

whence we obtain

$$\sum_{i=1}^{\infty} \frac{1}{(2i-1)^2} = \frac{\pi^2}{8}, \tag{2.134}$$

in agreement with eqn (2.71).

For other summations using Parseval's theorem, see Exercises 2.8 and 2.9.

2.11 Filters

A filter is a device which lets through some things and stops other things from getting through. This is true for coffee filters, for electrical filters and, in fact, for all filters of any kind I have ever come across.

A filter for a Fourier series falls into the same category. It may be defined as one which lets through some harmonics and rejects others. A more general definition of the function of a filter is that it should act upon the magnitude and relative phase of each harmonic.

Perhaps the simplest filter is the one which lets through unchanged all the harmonics up to N and cuts off all harmonics above that. We have, in fact, met such filters although we have not called them so. Let us look, for example, at Fig. 1.7. If we let the rectangular function of Fig. 1.7a through such a filter, we get f_1 of Fig. 1.7b for $N = 1$, f_2 of Fig. 1.7d for $N = 2$, f_3 of Fig. 1.7f for $N = 3$, etc. Thus, approximating a function by its first N Fourier components leads to the same result as sending the function through this filter.

Another, equally simple, filter would just let through the Nth harmonic and none of the others. For illustration we may look again at Fig. 1.7. If we let the pulse function of Fig. 1.7a through such a filter, we get f_1 of Fig. 1.7b for $N = 1$, p_2 of Fig. 1.7c for $N = 2$, p_3 of Fig. 1.7e for $N = 3$, etc.

The filter mentioned above belongs to a more general class of filters known as band pass filters. They will let through the harmonics between N_1 and N_2 and reject the others. What will happen if we apply such a

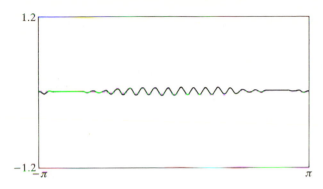

Fig. 2.28 The function of Fig. 1.7a after passing through a filter which cuts out all the components except those between 20 and 24

filter to the function shown in Fig. 1.7a? Nothing particularly interesting. For $N_1 = 20$ and $N_2 = 24$ we obtain the function shown in Fig. 2.28. As one would expect, it shows a band of frequencies.

We would fare similarly if we applied this kind of filter to any of the other functions I mentioned in this course, with the exception of the square wave modulated sinusoidal defined by eqn (2.94). If you remember, we used there the common-sense argument (supported by some mathematics as well) that the most important harmonics must be around the 100th harmonic. Just to check this statement, let us plot the function obtained by summing up all the components between 95 and 103. As may be seen in Fig. 2.29a, it resembles a sinusoidal and in the $t > 0.5T$ region it declines. We would of course obtain a better approximation by increasing the band of our bandpass filter. If we let through all the harmonics between 79 and 119, for example, we obtain the curve shown in Fig. 2.29b, and this is not very far from that shown in Fig. 2.13c, which is made up of the first 200 harmonics.

The point I am trying to make is that some desirable functions may essentially be built up by a band of frequencies in which case the application of a band pass filter causes little harm (in practice of course we use band pass filters because we do want to suppress frequencies outside a certain band).

The filters we have investigated so far have either let through a harmonic or completely rejected it. We shall now look at a filter, known as the "Hamming window" in signal analysis, which interferes with the amplitudes of the harmonics. The filter is defined by the relationship

$$F_k = \begin{array}{cc} 0.54 + 0.46 \cos(\pi k/N) & |k| \leqslant N \\ 0 & \text{if} \quad |k| > N \end{array} \qquad (2.135)$$

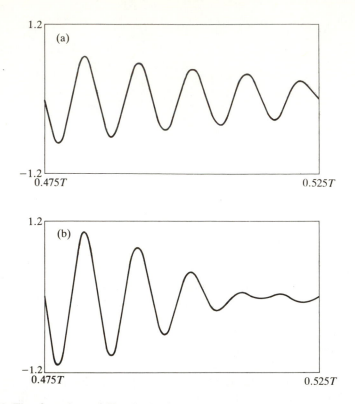

Fig. 2.29 The function of Fig. 2.12 after passing through a filter which cuts out all components except those between (a) 95 and 103, (b) 79 and 119

If the function $f(t)$ before the filter is represented by its Fourier series as

$$f(t) = \sum_{k=-\infty}^{\infty} c_k \exp\left(j\frac{2\pi kt}{T}\right), \tag{2.136}$$

then the new function, $f_f(t)$, after the filter, will take the form

$$f_f(t) = \sum_{k=-\infty}^{\infty} F_k c_k \exp\left(j\frac{2\pi kt}{T}\right), \tag{2.137}$$

i.e. the coefficient of each harmonic will be multiplied by F_k.

The variation of F_k as a function of k (only for positive values of k) is shown in Fig. 2.30 for $N = 17$. For low harmonics F_k is close to unity but it declines considerably for the higher harmonics. The 17th harmonic will only be 8% of that in the original function.

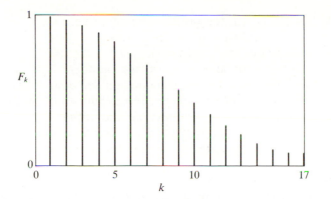

Fig. 2.30 The pass band characteristic of the filter defined by eqn (2.135)

What will be the effect of such a filter upon one of our functions introduced earlier? Let us choose again the function of Fig. 1.7a and plot the new function (let us call it f_{f17}) obtained for $N = 17$ after the filter shown in Fig. 2.31. We can now compare this curve with that of Fig. 1.7m which is the approximation given without the filter. You should immediately notice something remarkable. We have got rid of the Gibbs phenomenon. The ripples which have plagued us up to now have simply gone. We have an eminently smooth curve. Is this a matter for rejoicing? Is f_{f17} better than f_{17}? Better in what sense? The new function obtained with the Hamming window is not the optimum one, in the sense that it does not minimize the mean square error. The main disadvantage of the new function is that it rises and decays much more slowly than the old one. So it is a compromise. If you prefer aesthetics to utilitarianism then

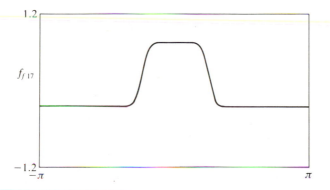

Fig. 2.31 The function shown in Fig. 1.7m after passing through the filter defined by eqn (2.135)

you should opt for f_{f17} (in the practical world of hardware, where these functions represent time-varying voltages, such smoothing is possible, but it needs the insertion of an actual filter).

There is, of course, nothing magic in the number 17. The Hamming window will get rid of the Gibbs phenomenon for other values of N as well. Its use is actually not limited to a rectangular function such as has been our chosen example, it will smooth all kind of functions in which the Fourier series has overshoots. And the Hamming window is not the only filter which can remove the ripples. Others are also possible, leading to different compromises.

There is no reason, of course, why a filter should affect only the amplitude of the Fourier components, it could just as well affect the phases. We shall illustrate such a filter using again the function presented in Fig. 1.7a. Its Fourier series without the constant term is given by eqn (2.127) for $\tau = T/4$, yielding

$$f(x) = \frac{2}{\pi} \sum_{k=1}^{\infty} \frac{1}{k} \sin \frac{\pi k}{4} \cos kx. \tag{2.138}$$

Now one way of affecting the phase of this function would be to add to each term a progressive phase $k\varphi$, in which case eqn (2.138) modifies to

$$f(x) = \frac{2}{\pi} \sum_{k=1}^{\infty} \frac{1}{k} \sin \frac{\pi k}{4} \cos[k(x + \varphi)]. \tag{2.139}$$

It may now be seen that adding such progressive phase means no more than a shift of the whole pattern without any distortion. A constant phase added to each harmonic would, on the other hand, entail a different shift for each harmonic (smaller relative shift for higher harmonics) and hence it would affect the shape of the waveform. With the constant phase added, the Fourier series takes the form

$$f(x) = \frac{2}{\pi} \sum \frac{1}{k} \sin \frac{\pi k}{4} \cos\left[k\left(x + \frac{\varphi_1}{k}\right)\right] \tag{2.140}$$

The first 17 components (called this time $f_{\varphi 17}$) are plotted in Fig. 2.32a and b for $\varphi = 0.2$ and 0.4 respectively. Quite obviously, as the components are added in the wrong phase the waveform gets distorted.

Different functions have, of course, different sensitivities to the relative phases, so the values of 0.2 and 0.4 given above are by no means typical. To get a little more 'feel' for the effect of relative phases, attempt Exercise 2.20.

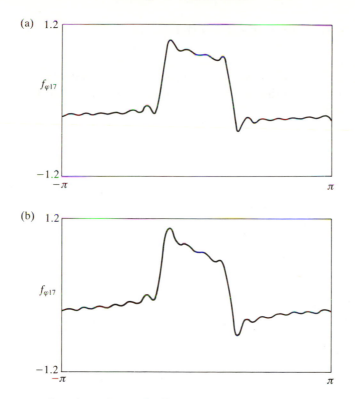

Fig. 2.32 The function shown in Fig. 1.7m after passing through a filter which shifts the phase of each component by (a) 0.2 radian and (b) 0.4 radian.

Exercises

2.1. A periodic function of period 2π is defined in the interval $-\pi, \pi$ as follows:

$$f(x) = \begin{matrix} 0 \\ \cosh x \end{matrix} \quad \text{if} \quad \begin{matrix} -\pi < x < 0 \\ 0 < x < \pi \end{matrix}. \qquad (2.141)$$

(i) Find the Fourier series.
(ii) Show by considering the Fourier series at a certain point that

$$\sum_{k=1}^{\infty} \frac{(-1)^k}{1+k^2} = \frac{1}{2}\left(\frac{\pi}{\sinh \pi} - 1\right). \qquad (2.142)$$

2.2. A function is defined in the interval 0, 1 as

$$f(t) = \sinh t. \qquad (2.143)$$

Sketch the function as it will be extended for
  (i) a full-range Fourier series, i.e. expressed as a function of $\cos 2\pi kt$ and $\sin 2\pi kt$;
 (ii) a half-range sine series, i.e. expressed as a function of $\sin \pi kt$;
(iii) a half-range cosine series, i.e. expressed as a function of $\cos \pi kt$.
  Find the fourier series for all three cases.
2.3. A periodic function of period 2π is defined in the interval $-\pi, \pi$ as follows:

$$f(x) = \begin{matrix} -\cos(x/2) \\ \cos(x/2) \end{matrix} \quad \text{if} \quad \begin{matrix} -\pi < x < 0 \\ \pi < x < 2\pi \end{matrix}. \tag{2.144}$$

Find the Fourier series.
2.4. A function is defined in the interval 0.1 as

$$f(t) = \exp(-2t^2). \tag{2.145}$$

  Find numerically the first three Fourier components for a half-range cosine series. Plot the original function and the approximate series obtained in the same diagram.
2.5. In the function defined in eqn (2.94), it was assumed that $\omega_0 = 100\omega$. The corresponding Fourier series was given by eqn (2.96). Determine the Fourier series when $\omega_0 = 99.9\omega$.
2.6. Find the infinite sum

$$\sum_{k=1}^{\infty} \frac{1}{1+k^2} \tag{2.146}$$

with the aid of the Fourier series given by eqn (2.64).
2.7. Find the infinite sum

$$\sum_{i=1}^{\infty} \frac{(-1)^{i-1}}{(2i-1)^3} \tag{2.147}$$

with the aid of the Fourier series given by eqn (2.112).
2.8. Show with the aid of Parseval's theorem that

$$\sum_{k=1}^{\infty} \frac{1}{k^4} = \frac{\pi^4}{90}. \tag{2.148}$$

Use the Fourier series given by eqn (2.109).
2.9. Show with the aid of Parseval's theorem that

$$\sum_{i=1}^{\infty} \frac{1}{(2i-1)^6} = \frac{\pi^6}{960}. \tag{2.149}$$

Use the Fourier series given by eqn (2.109).
2.10. Confirm that eqn (2.90) gives the Fourier series of the function

shown in Fig. 2.10. Program the series on a computer and plot it for various values of N. Try to predict how many terms you need for a 'reasonable approximation'. Can you say what you mean by a 'reasonable approximation'?

2.11. Confirm that eqn (2.93) gives the Fourier series of the rectified cosine function shown in Fig. 2.11b. Program the series on a computer and plot it for various values of N. Try to predict how many terms you need for a reasonable approximation.

2.12. Confirm that eqn (2.105) gives the Fourier series for the function shown in Fig. 2.12b.

2.13. Confirm that eqn (2.107) gives the Fourier series for the function shown in Fig. 2.12c.

2.14. Confirm that eqn (2.109) gives the Fourier series for the function shown in Fig. 2.12d.

2.15. Confirm that eqn (2.112) gives the Fourier series for the function shown in Fig. 2.18.

2.16. Confirm that eqn (2.96) gives the Fourier series of the function defined in eqn (2.94). (Hint: Care needs to be exercised in deriving the sine components.)

2.17. Apply the Hamming window (given by eqn (2.135)) to f_6 and f_9 plotted in Fig. 1.7j and Fig. 1.7k respectively. Plot the results.

2.18. $(x/2\pi)^2$ is plotted in Fig. 2.8 for the first 20 harmonics. Plot it again after applying the Hamming window to it.

2.19. Define a filter function by the following relations

$$F_k = \begin{matrix} 0.08 + 0.92 \cos^2(\pi k/2N) \\ 0 \end{matrix} \quad \text{if} \quad \begin{matrix} |k| \leq N \\ |k| > N \end{matrix}. \qquad (2.150)$$

Apply this window to f_{17}, the Fourier series plotted in Fig. 1.7m. Compare the function obtained with that shown in Fig. 2.31.

2.20. For $\varphi = 0$, the Fourier series

$$f(x) = \frac{2}{\pi} \sum_{k=1}^{\infty} \frac{1}{k} \sin \frac{\pi k}{4} \cos k(x + k\varphi) \qquad (2.151)$$

is that of the function shown in Fig. 1.7a. For φ different from zero the harmonics are shifted relative to each other, introducing some distortion. Plot the resulting curves obtained by taking the first 20 harmonics and choosing $\varphi = 0.005$, 0.01 and 0.02.

3
Forced solutions of ordinary differential equations with periodic excitation

3.1 Introduction

IT IS not the aim of this chapter to review the solution of ordinary differential equations in general. If the reader is familiar with the various solutions, so much better, but such knowledge will not be necessary for following the subsequent argument. I shall make an attempt to explain everything, and explain everything slowly.

In the first five sections we shall be concerned with a rather specific differential equation, one which has constant coefficients and which is driven by a periodic function. A few examples will make clear what I mean. Let us see first the differential equation:

$$y' = \sin x \qquad (3.1)$$

This is a differential equation all right, because the derivative $y' = \mathrm{d}y/\mathrm{d}x$ figures in it, and it is first-order because y' is the highest derivative. It is also a linear differential equation with constant coefficients, because y' is multiplied by 1, which is a constant. On the right-hand side we have $\sin x,$ which is a periodic function. To be 'driven' or 'excited' by a periodic function means that there is a 'primary' variation to which the unknown function responds. If the excitation is periodic, the response will also be periodic. The response, we say, is forced.

Now let us return to our differential equation (3.1). We don't need to ponder too much on how to solve it. We shall simply integrate eqn (3.1) and obtain,

$$y = -\cos x. \qquad (3.2)$$

This is simple indeed. You may object at this point and say that this is not general enough. You might like to see here a constant as well. We don't have it because we don't need it. We want only the forced response.

We shall go rather slowly, starting first with a linear differential equation of the first order then proceeding to the second order and finally to arbitrary order, all under conditions of forced response.

But what has this got to do with Fourier series? The point is that an arbitrary periodic excitation may be represented by its Fourier series. The solution, the response to the excitation, will be another Fourier series. Our concern is the relationship between the two series.

In what branches of science will such relationships be of importance? Whenever there are periodic excitations. Mechanical vibrations in mechanical engineering are good examples, but perhaps the most important single field is signal analysis in electrical engineering. The question asked is then what is the output of a system for a specified input. In Section 3.5, answers in a few specific cases will be given.

3.2   First-order differential equations

We shall now take a slightly more complicated differential equation:

$$y' + y = \sin x. \tag{3.3}$$

This may again be recognized as a linear differential equation with constant coefficients driven by a periodic function. How shall we solve it? To find a solution is to find a function $y = f(x)$ which will satisfy eqn (3.3). How to find that function? By intuition. We may expect that the response to a periodic excitation will also be periodic and will have the same period. Now the driving term is a function $\sin x$ which has a period 2π. The periodic functions with a period 2π are $\cos kx$ and $\sin kx$, provided k is an integer. But remember that our differential equation is linear. There will be no values of k higher than 1 because there is no mechanism capable of producing higher harmonics.† Consequently, we can have terms only in $\cos x$ and $\sin x$.

So how can we find the solution? It is easy with differential equations. If we have some inkling which way the solution lies we can always assume a trial solution, and we can learn soon enough whether it is correct or not. The logical thing is to assume the solution as a linear combination of cosine and sine functions in the form

$$y = a \cos x + b \sin x, \tag{3.4}$$

where a and b are unknown coefficients.

The rest is mathematical technique. I know this is easy, but doing it the first time I want to go ahead slowly and do all the steps. Substituting eqn

---

† There is, though, such a mechanism in a non-linear differential equation. Take one, for example, which contains a term y^2 and which is driven by $\cos x$ or $\sin x$. Then a solution like $\cos x$ or $\sin x$ (or their combination) could not possibly work, because when they are squared they give rise to $\cos 2x$ and $\sin 2x$ terms. These problems will be discussed briefly in Section 3.6.

(3.4) into (3.3) we obtain

$$b \cos x - a \sin x + b \sin x + a \cos x = \sin x. \tag{3.5}$$

Equating the coefficients of $\sin x$ and $\cos x$ on both sides, we obtain the algebraic equations

$$b + a = 0 \tag{3.6}$$

and

$$-a + b = 1, \tag{3.7}$$

which have the solution

$$a = -\tfrac{1}{2}, \qquad b = \tfrac{1}{2}, \tag{3.8}$$

yielding for the solution of the differential equation

$$y = \tfrac{1}{2}(\sin x - \cos x). \tag{3.9}$$

  Those in the know will say that this is not the most general solution, that the general solution is obtained by adding the solution of the homogeneous differential equation

$$y' + y = 0 \tag{3.10}$$

to the solution we have obtained in eqn (3.9). This is true, but, as I mentioned before, for the purpose of this chapter the general solution is irrelevant. We are only looking for the 'forced' solution. If x represents time then we can say that we are concerned with the stationary solution—with the solution that occurs when x is large enough that the transients will have died away. The solution at $x = 0$ is of no interest. We are not concerned with the initial conditions.

3.3   Second-order differential equations

Let us progress now from a first-order to a second-order differential equation of the form

$$y'' + 2\zeta y' + \omega_0^2 y = a \cos x + b \sin x, \tag{3.11}$$

where ζ and ω_0^2 are constants (we have taken the constants in this rather peculiar form because it will make our later equations somewhat simpler).

  The solution may again be assumed as a combination of sine and cosine functions:

$$y = A \cos x + B \sin x. \tag{3.12}$$

Substituting eqn (3.12) into (3.11) we obtain

$$-A \cos x - B \sin x + 2\zeta(-A \sin x + B \cos x)$$
$$+ \omega_0^2(A \cos x + B \sin x) = a \cos x + b \sin x. \quad (3.13)$$

In order to satisfy eqn (3.13), the coefficients of $\cos x$ and $\sin x$ must be the same on both sides, leading to the equations for the unknown coefficients A and B as follows:

$$-A + 2\zeta B + \omega_0^2 A = a \quad (3.14)$$

and

$$-B - 2\zeta A + \omega_0^2 B = b. \quad (3.15)$$

We have two equations, two unknowns: they can be easily solved to give

$$A = \frac{a(\omega_0^2 - 1) - 2\zeta b}{\omega_0^2 + 4\zeta^2}, \qquad B = \frac{2\zeta a + b(\omega_0^2 - 1)}{\omega_0^2 + 4\zeta^2}. \quad (3.16)$$

We can conclude† that with the aid of A and B we now have a solution in terms of another periodic function which has a different amplitude and phase from the driving function.

If the driving periodic function is not a simple sinusoidal but has a number of higher harmonics as well, the solution is still of the same type. We just need to add the effect of each harmonic.

3.4 Higher-order differential equations

We are now ready to tackle the general nth order differential equation driven by a periodic function given in the form of a Fourier series. The general form is

$$\sum_{i=0}^{n} g_i y^{(i)} = \frac{a_0}{2} + \sum_{k=1}^{\infty} (a_k \cos kx + b_k \sin kx). \quad (3.17)$$

On the left-hand side all the g_i are constants, $y^{(i)}$ is the ith derivative with respect to x, and $y^{(0)} = y$. On the right-hand side we have a general periodic excitation in the form of a Fourier series. Since the equation is linear it is sufficient to find the solution for one value of k and this will be

---

† For the purists it must be noted here that eqn (3.16) gives no solution (both A and B are equal to zero) when $\omega_0 = 1$ and $\zeta = 0$. In that case the proper solution is not in the form of a periodic function. All elementary textbooks of applied mathematics discuss this case (and some more complicated ones belonging to the same class) with gusto, but I have never come across a problem in engineering where such non-periodic solutions described the relevant physical phenomena, so I shall just disregard that possibility.

valid for all values of k. So let us attempt the solution in the form

$$y = A_k \cos kx + B_k \sin kx. \tag{3.18}$$

Then

$$\sum_{i=0}^{n} g_i y^{(i)} = \sum_{i=0}^{n} g_i \left(A_k \frac{d}{dx^i} \cos kx + B_k \frac{d}{dx^i} \sin kx \right). \tag{3.19}$$

Note that the derivatives of $\cos kx$ and $\sin kx$ will be just some other combination of $\cos kx$ and $\sin kx$, so we can again equate the coefficients of $\cos kx$ and $\sin kx$ on both sides of eqn (3.19), yielding two algebraic equations for A_k and B_k.

This is as far as I intend to go in generality and this can be the end of the matter for those who have a natural dislike of complex numbers. However, for those willing to tackle complex numbers, a great reward is coming. The technique of solving eqn (3.17) will be considerably simplified.

Let us first rewrite the Fourier series in its complex form as given in eqn (2.122) with t as an independent variable. Equation (3.17) may then be recast in the form

$$\sum_{i=0}^{n} g_i y^{(i)} = \sum_{k=-\infty}^{\infty} c_k \exp\left(j \frac{2\pi kt}{T} \right), \tag{3.20}$$

where $y^{(i)}$ now means differentiation with respect to t.

We shall attempt the solution in the form

$$y = \sum_{k=-\infty}^{\infty} C_k \exp\left(j \frac{2\pi kt}{T} \right). \tag{3.21}$$

differentiating eqn (3.21) i times gives

$$y^{(i)} = \sum C_k \left(j \frac{2\pi k}{T} \right)^i \exp\left(j \frac{2\pi kt}{T} \right). \tag{3.22}$$

Substituting eqn (3.22) into eqn (3.20), we obtain for any given value of k,

$$\sum_{i=0}^{n} g_i \left(j \frac{2\pi k}{T} \right)^i C_k \exp\left(j \frac{2\pi kt}{T} \right) = c_k \exp\left(j \frac{2\pi kt}{T} \right), \tag{3.23}$$

whence the solution for C_k emerges immediately as

$$C_k = \frac{c_k}{\sum_{i=0}^{n} g_i \left(j \frac{2\pi k}{T} \right)^i}. \tag{3.24}$$

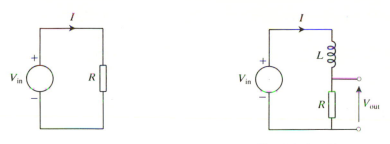

Fig. 3.1 A resistive circuit   **Fig. 3.2** An *RL* circuit

3.5   Examples from electrical engineering

1. *A resistive circuit.* Electrical engineering is of course the ideal testing ground for solving ordinary differential equations. The exception is this first example in which there are no differentials at all. The aim is here to work out power.

We shall assume that the voltage is known as a function of time and we wish to find the current in the circuit of Fig. 3.1. This is a rather simple problem. All we need to do is to apply Ohm's law. For a voltage $V(t)$ the current is

$$I(t) = \frac{V(t)}{R}. \tag{3.25}$$

If now the voltage is periodic given in the form of a Fourier series

$$V(t) = \sum_{k=-\infty}^{\infty} c_k \exp\left(j\frac{2\pi kt}{T}\right) \tag{3.26}$$

then we obtain the current by simply dividing each Fourier coefficient by the value of the resistance, R. Let us now find the power. It is defined as

$$P = \frac{1}{T}\int_0^T V(t)\, I(t)\, dt = \frac{1}{R}\frac{1}{T}\int_0^T V^2(t)\, dt. \tag{3.27}$$

Thus, integration in time is one way of finding the power. We can, however, use the fact that $V(t)$ is given by its Fourier series, and then using Parseval's theorem (eqn (2.130)) we may find the power in the alternative form

$$P = \frac{1}{R}\sum_{k=-\infty}^{\infty} |c_k|^2 = \frac{1}{R}\left[\frac{a_0^2}{4} + \frac{1}{2}\sum_{k=1}^{\infty}(a_k^2 + b_k^2)\right]. \tag{3.28}$$

If you think about it, it makes good physical sense. For a d.c. voltage of V_0 the power absorbed is V_0^2/R. For a sinusoidal input $V_1 \cos \omega t$, the power is known to be $V_1^2/2R$. If we add to the voltage the second harmonic $V_2 \cos 2\omega t$ the power due to that is $V_2^2/2R$, etc., ending up with

$$P = \frac{1}{R}\left[V_0^2 + \frac{1}{2}\sum V_n^2 \right], \tag{3.29}$$

which agrees with eqn (3.28), considering that

$$a_0 = 2V_0, \qquad a_k = V_k, \qquad b_k = 0. \tag{3.30}$$

2. *An RL circuit.* We shall now look at the electric circuit of Fig. 3.2. This is the kind of example for differential equations in which the physics and the mathematics mutually reinforce each other. The mathematics is not on its own. We have some expectations from the physics which the mathematics must provide. We know from circuit theory that the input voltage V_{in} is related to the current by the differential equation

$$L\frac{dI}{dt} + RI = V_{in}, \tag{3.31}$$

where I is the current.

You are familiar with the solution of this problem for a sinusoidal input. Let us now choose for our input voltage a more complicated waveform, namely a square wave of amplitude V_s as shown in Fig. 3.3. The Fourier coefficients in the exponential representation may be calculated from eqn (2.125) in the form

$$c_k = 2V_s\frac{\sin(\pi k/2)}{\pi k/2}, \qquad c_0 = 0. \tag{3.32}$$

The problem is now to find the waveform of the output function, $V_{out} = RI$.

In accordance with the treatment given in the previous section, the trial

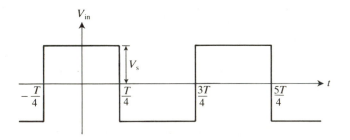

Fig. 3.3 A square wave input voltage

solution is

$$I = \sum_{k=-\infty}^{\infty} C_k \exp\left(j\frac{2\pi kt}{T}\right) \tag{3.33}$$

and the formula for C_k is

$$C_k = \frac{c_k}{R + jk\omega L}, \qquad \omega = \frac{2\pi}{T} \tag{3.34}$$

Note that eqn (3.34) follows from eqn (3.24) by assigning

$$g_0 = R \qquad \text{and} \qquad g_1 = L. \tag{3.35}$$

The output voltage is then given by

$$V_{\text{out}} = RI = \sum_{k=-\infty}^{\infty} \frac{c_k \exp(jk\omega t)}{1 + jk\eta}, \qquad \eta = \frac{\omega L}{R}. \tag{3.36}$$

If we wish we can, of course, return to the trigonometric form by using the relationship between the coefficients (eqn (2.126)) in the two representations, yielding

$$V_{\text{out}} = 2 \sum_{k=1}^{\infty} c_k \left[\frac{1}{1 + k^2\eta^2} \cos k\omega t + \frac{k\eta}{1 + k^2\eta^2} \sin k\omega t \right]. \tag{3.37}$$

The only parameter is now η. When it is small, the output waveform is rather similar to the input waveform, as may be seen in Fig. 3.4a and b for $\eta = 0.1$ and 0.2 respectively. In both cases the transition from negative to positive values is quite smooth (there is some discrimination against higher harmonics) although the Gibbs phenomenon has not been completely eliminated. As η increases, the output waveform changes further, as shown in Figs. 3.4c, d, and e for $\eta = 0.5$, 2 and 10 respectively. Eventually it becomes a pair of straight lines, i.e. the integral of the input waveform. This is not particularly surprising. If η is large it means that the reactance is large in comparison with the resistance, i.e. the resistance may be neglected in eqn (3.31), and we are left with a simple integral relationship between the current (proportional to output voltage) and the input voltage.

3. *An RLC circuit.* For our next example we choose the circuit shown in Fig. 3.5, in which a current generator feeds an inductor, a capacitor, and a resistance in parallel. It is now the input current which is given in the form of a Fourier series. The aim is to determine the voltage across the parallel combination. According to the rules of circuit theory, we may write

$$\frac{1}{C} \int I_C \, dt = L \frac{dI_L}{dt} = RI_R = V_{\text{out}} \tag{3.38}$$

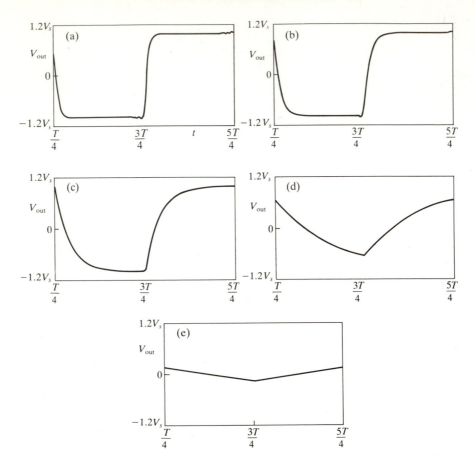

Fig. 3.4 The output waveform as given by eqn (3.37) for the input waveform of Fig. 3.3 for $N = 40$. (a) $\eta = 0.1$, (b) $\eta = 0.2$, (c) $\eta = 0.5$, (d) $\eta = 2$, and (e) $\eta = 10$

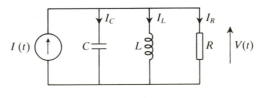

Fig. 3.5 An *RLC* circuit

and

$$I = I_C + I_L + I_R. \tag{3.39}$$

We may now express all the terms on the right-hand side with the aid of V as follows

$$I = C\frac{dV}{dt} + \frac{V}{R} + \frac{1}{L}\int V \, dt. \tag{3.40}$$

The current is now assumed to be given as

$$I = \sum_{k=-\infty}^{\infty} c_k \exp\left(j\frac{2\pi kt}{T}\right) \tag{3.41}$$

and we wish to find

$$V = \sum_{k=-\infty}^{\infty} C_k \exp\left(j\frac{2\pi kt}{T}\right), \tag{3.42}$$

where C_k are the coefficients to be determined.

Substituting eqns (3.41) and (3.42) into (3.40) we find the following equation for each value of k:

$$c_k = \left(jk\omega C + \frac{1}{R} + \frac{1}{jk\omega L}\right)C_k \tag{3.43}$$

Expressing C_k by eqn (3.43) and substituting it into eqn (3.42), we have our formula for the voltage, $V(t)$. If we want to plot any curves we need, of course, give the waveform of the input current. If we choose (as in the previous example) a square wave function of amplitude I_s whose coefficients are given in the same form as in eqn (3.32) and we do the necessary algebra to convert our expression into a trigonometric form, we obtain at the end

$$V = I_s R \frac{4}{\pi} \sum_{k=1}^{\infty} \frac{(-1)^{i-1}}{2i-1} \frac{1}{1 + \left[(2i-1)\alpha - \frac{1}{(2i-1)\zeta}\right]^2}$$

$$\times \left\{ \cos(2i-1)\omega t + \left[(2i-1)\alpha - \frac{1}{(2i-1)\zeta}\right]\sin(2i-1)\omega t \right\}, \tag{3.44}$$

where $\alpha = \omega RC$ and $\zeta = \omega L/R$.

It may be seen from eqn (3.44) that when $\alpha = 0$ and $\zeta = \infty$, i.e. when neither the capacitance nor the inductance are present, we obtain the original square wave input. When α and ζ are finite, we may obtain a wide variety of waveforms. We shall investigate only some of them, those which are affected by the resonance property of the circuit.

It follows from circuit theory that the nth harmonic is at resonance

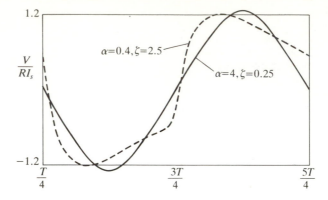

Fig. 3.6 Output voltage when the circuit of Fig. 3.5 is tuned to resonance for the fundamental component: (– – –) weak resonance: (——) strong resonance

when

$$\alpha\zeta = \frac{1}{n^2} \quad \text{or} \quad LC = \frac{1}{(n\omega)^2}. \tag{3.45}$$

What does it mean to say that it is at resonance? It means that there is discrimination in favour of the harmonic at resonance against the others, which are suppressed by varying amounts. The amount of suppression depends, of course, on the strength of the resonance. The resonance is strong when the impedances of the capacitor and of the inductor are small relative to the resistance, i.e. when

$$\alpha \gg 1 \quad \text{and} \quad \zeta \ll 1. \tag{3.46}$$

In the following numerical examples we shall first assume that the fundamental component is at resonance, i.e. $\alpha\zeta = 1$ and take $\alpha = 4$, $\zeta = 0.25$ and $\alpha = 0.4$, $\zeta = 2.5$. The corresponding curves are shown in Fig. 3.6 with solid and dotted lines respectively. When the resonance is weak (dotted lines) the other components, beside the resonant, are not sufficiently suppressed, with the result that there is still some intimation of the square wave in the waveform obtained. However, the curve for $\alpha = 4$, $\zeta = 0.25$ (solid lines) seems to indicate that the resonance is strong enough that only the fundamental component is left. We can also say that in this case the resonant circuit behaves as a filter which lets through only the fundamental component.

Next we tune our circuit to 5th harmonic resonance, which requires that $\alpha\zeta = 0.04$. For a rather weak resonance ($\alpha = 0.4$, $\zeta = 0.1$—Fig. 3.7a), there are other harmonics present as well beside the 5th harmonic.

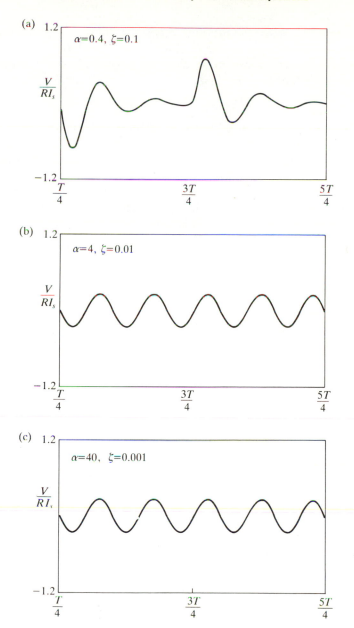

Fig. 3.7 Output voltage when the circuit of Fig. 3.5 is tuned to resonance for the fifth harmonic. Parts (a), (b), and (c) correspond to weak, stronger, and strong resonance

The strong influence of the 5th harmonics can, however, be discerned by noticing that there are five peaks. For $\alpha = 4$, $\zeta = 0.01$ (Fig. 3.7b), the other harmonics are quite small; and for $\alpha = 40$, $\zeta = 0.001$ (Fig. 3.7c), they seem to be entirely negligible.

3.6   Solution of a non-linear differential equation

In this section we shall look at a particular example of a non-linear ordinary differential equation which can be solved approximately using Fourier techniques. The example is taken from real life. The differential equation does arise in a branch of optics. The periodic excitation is due to the interference of two plane waves which may be described as

$$I = 1 + m \cos x, \tag{3.47}$$

where I is normalized intensity, m is the depth of modulation (assumed to be small) and x is the spatial coordinate.

The aim is to find the response of the non-linear system. Since the excitation is periodic, the response must also be periodic. But, and this is where the treatment in the present section differs from those in the previous ones, the response to a sinusoidal excitation will not be purely sinusoidal. Owing to the non-linear character of the differential equation, the response will contain higher harmonics as well.

From now on I shall say no more about the physics (it is fairly complicated, so it is not worth including it here) and concentrate on the mathematics. The differential equations to be solved are as follows,

$$y' = z \tag{3.48}$$

and

$$z' = z^2 + y - (1 + m \cos x). \tag{3.49}$$

As may be seen, we have two independent variables y and z and the non-linear term is z^2.

We shall now attempt a solution in the form of a finite Fourier series

$$y = y_0 + y_1 \cos x + y_2 \cos 2x, \tag{3.50}$$

where y_0, y_1 and y_2 are unknown constants. It follows that

$$z = -y_1 \sin x - 2y_2 \sin 2x,$$
$$z' = -y_1 \cos x - 4y_2 \cos 2x, \tag{3.51}$$

and

$$z^2 = y_1^2 \sin^2 x + 4y_2^2 \sin^2 2x + 4y_1 y_2 \sin x \sin 2x$$

$$= \frac{y_1^2}{2}(1 - \cos 2x) + 2y_2^2(1 - \cos 4x) + 2y_1 y_2(\cos x - \cos 3x)$$

$$\cong \frac{y_1^2}{2} + 2y_2^2 + 2y_1 y_2 \cos x - \frac{y_1^2}{2}\cos 2x, \tag{3.52}$$

where we have disregarded the $\cos 3x$ and $\cos 4x$ terms.

Substituting eqns (3.50) and (3.51) into (3.49) we obtain

$$-y_1 \cos x - 4y_2 \cos 2x = \frac{y_1^2}{2} + 2y_2^2 + 2y_1 y_2 \cos x$$

$$- \frac{y_1^2}{2}\cos 2x + y_0 + y_1 \cos x + y_2 \cos 2x$$

$$- 1 - m \cos x. \tag{3.53}$$

The above equation is equivalent to three equations which we may obtain by equating the constants, and the coefficients of $\cos x$ and $\cos 2x$, yielding

$$0 = \frac{y_1^2}{2} + 2y_2^2 + y_0 - 1 \tag{3.54}$$

$$-y_1 = 2y_1 y_2 + y_1 - m \tag{3.55}$$

$$-4y_2 = -\frac{y_1^2}{2} + y_2. \tag{3.56}$$

Now remember that m is small in comparison with unity, i.e. the excitation is a weak sinusoidal on the top of a constant term. Hence one may expect that in the solution too y_1 is much less than y_0 and, similarly, y_2 is much less than y_1.

If this is true then y_1^2 in eqn (3.54) is of second order and y_2^2 is of fourth order. Relative to y_0 (which is of zero order) they are both negligible. Hence, in this approximation, our first solution is

$$y_0 = 1. \tag{3.57}$$

Looking at eqn (3.55), we see that m and y_1 are of first order and $y_1 y_2$ is of the third order. Thus $y_1 y_2$ may be neglected and eqn (3.55) may be solved to yield

$$y_1 = \frac{m}{2}. \tag{3.58}$$

There are no terms to neglect in eqn (3.56). All three terms are of the second order. Solving for y_2 we obtain

$$y_2 = \frac{y_1^2}{10} = \frac{m^2}{40}.$$ (3.59)

We have now found an approximation for all the unknowns, but we need not stop at this stage, we can find higher-order approximations.

Remember that in the first round we neglected $y_1^2/2$ in eqn (3.54). But having found y_1 in the form of eqn (3.58) we may now substitute it into eqn (3.54) and find the higher-order approximation for y_0 in the form

$$y_0 = 1 - \frac{y_1^2}{2} = 1 - \frac{m^2}{8}.$$ (3.60)

Similarly, we may claim that having worked out y_1 and y_2 we can put $y_1 y_2 = m^3/80$ back into eqn (3.55) and find the higher-order approximation for y_1 in the form

$$y_1 = \frac{m}{2} + \frac{m^3}{80} = \frac{m}{2}\left(1 + \frac{m^2}{40}\right).$$ (3.61)

Obviously, the smaller m is the better is the approximation. If m is equal to (say) 0.2 then the correction term in eqn (3.60) is 0.005. Thus, the zero-order approximation is correct to about 0.5%. That is not bad. But what happens when m is considerably greater? What if m approaches unity: how could we improve the approximation? The answer is that we need to take more terms into account in the attempted solution, say up to $\cos Nx$. Unfortunately, even if N is as low as 4 the amount of mathematical labour involved in trying to match the coefficients up to $\cos 4x$ is quite significant. And of course I have chosen a relatively simple differential equation in which the non-linearity is relatively small and in which only cosine terms need to be considered. In the general case sine terms will be present as well, immediately doubling the number of unknown coefficients.

The conclusion is that the solution of non-linear differential equations with periodic excitation is difficult. It is a fairly new discipline to which a lot of attention is being devoted, particularly by numerical analysts, but for which no satisfactory solution exists for the moment.

Exercises

3.1. Find the forced response $V_{out}(t)/I_s R$ in the form of a Fourier series for the triangular input current defined as

$$f(t) = \begin{matrix} 2I_s t/T \\ 2I_s(1 - t/T) \end{matrix} \quad \text{for} \quad \begin{matrix} 0 < t < T/2 \\ T/2 < t < T \end{matrix}$$ (3.62)

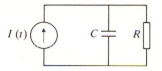

Fig. 3.8 A parallel *RC* circuit fed by a current source

applied to the circuit shown in Fig. 3.8. Plot the waveform for $\omega RC = 0.01$, 1 and 100 where $\omega = 2\pi/T$.

3.2. Find the forced response $V_{out}(t)$ in the form of a Fourier series for a triangular input voltage (replace I_s by V_s in eqn (3.62)) applied to the circuit shown in Fig. 3.9. Plot the waveform for $\omega RC = 0.01$, 10 and 100.

3.3. Find the forced response $V_{out}(t)$ in the form of a Fourier series for a square wave voltage input (Fourier coefficients given by eqn (3.32)) applied to the series resonant circuit shown in Fig. 3.10. Plot the waveform when the resonance condition $n^2\omega^2 LC = 1$ is satisfied for the fundamental component $(n = 1)$ and for the third harmonic $(n = 3)$, where $\omega = 2\pi/T$, and vary the strength of the resonance by considering the parameters $\alpha = \omega RC$ and $\zeta = \omega L/R$. Note that for strong resonance α must be small and ζ large.

3.4. Solve the non-linear differential equation (eqns (3.48) and (3.49)) for the impressed intensity $1 + m \cos x$ to third-order approximation, i.e. by assuming the solution in the form

$$y = y_0 + y_1 \cos x + y_2 \cos 2x + y_3 \cos 3x \qquad (3.63)$$

and then determining the unknown constants y_0, y_1, y_2 and y_3.

3.5. Solve the non-linear differential equations

$$y' = z \qquad (3.64)$$

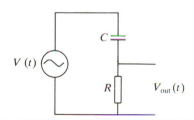

Fig. 3.9 A series *RC* circuit fed by a voltage source

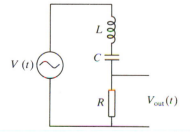

Fig. 3.10 A series resonant circuit fed by a voltage source

and

$$z' = z^2 + \frac{z}{y} + y - (1 + m \cos x),\qquad(3.65)$$

a more complicated version of eqns (3.48) and (3.49), up to second-order approximation. Hint: The solution is no longer symmetric; hence, instead of eqn (3.50), it should be attempted in the form

$$y = y_0 + \sum_{k=1}^{3} (y_{ck} \cos kx + y_{sk} \sin kx)\qquad(3.66)$$

4
Partial differential equations

4.1 Introduction

PARTIAL differential equations differ from ordinary differential equations by having derivatives with respect to more than one variable. The fact that one needs to differentiate once by one variable and then by another variable does not really tax one's mathematical ingenuity. If $z(x, y)$ is a function of the variables x and y then $\partial z/\partial x$ means differentiation by x while y is taken as a constant, and $\partial z/\partial y$ means differentiation by y while x is taken a constant. And of course we do the same thing if we have more variables or if we want to find higher partial derivatives.

In a partial differential equation of the nth order, some combination of the function and its partial derivatives (up to the nth) appear. A solution is obtained when we have managed to determine the unknown function. It is very easy to know whether we have found the right solution. The function obtained needs to be substituted into the differential equation. It is a solution if it satisfies the differential equation. For example,

$$z(x, y) = x + y \tag{4.1}$$

is a solution of the differential equation

$$\frac{\partial^2 z}{\partial x^2} + \frac{\partial^2 z}{\partial y^2} = 0. \tag{4.2}$$

Incidentally, in order to save space, from now on we shall adopt the subscript for denoting partial differentiation. Thus the above partial differential equation is written in the form

$$z_{xx} + z_{yy} = 0. \tag{4.3}$$

This is a fairly simple equation. But, you may ask, will all partial differential equations be that simple? Well, some partial differential equations are more complicated than others, but those included in the present course will certainly be simple. Will the boundary conditions be simple? Well, in one sense they are bound to be more complicated for partial differential equations: namely there will be conditions in more than one variable. Then will the solutions be very complicated? Well, in some cases they will look very complicated but, essentially, they will be put together from simple solutions. The aim is, of course, to investigate problems which can be solved by the application of a Fourier series. So once again you will only need to apply the rules you have learned.

If the partial differential equations are simple and their solutions are also relatively simple, does it mean that the problems are so much oversimplified that they lose all resemblance to reality? No, not at all. The problems are genuine, and if you solve them, plot them, and look at them, you will have gained an understanding of a number of physical processes and, at the same time, you will have had an appreciation of the power of Fourier's method.

We shall look at three sets of problems, namely at the vibrating string in Section 4.2, at heat conduction in Section 4.3, and at transmission line problems in Section 4.4.

4.2   The vibrating string

A string stretched between two points may well serve as an introduction to the study of wave propagation. It is sufficiently close to everyday experience that one can have an intuitive 'feel' for what is going to happen for given initial conditions.

First, of course, we need the relevant equation for a vibrating string. It is a standard piece of derivation which can be found in many textbooks. We shall do it here for completeness.

We start with an element of the string between x and $x + \Delta x$, as shown in Fig. 4.1. At the points x and $x + \Delta x$ it has transverse displacements of $y(x)$ and $y(x + \Delta x)$, and the corresponding slopes are

$$\tan \alpha = y_x(x) \qquad \text{and} \qquad \tan \beta = y_x(x + \Delta x) \qquad (4.4)$$

respectively.

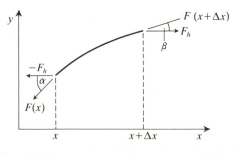

Fig. 4.1 Forces on an elementary string

Fig. 4.2 Vector triangles showing the forces at (a) x and (b) $x + dx$

The horizontal components of the force, F_h, are assumed to be equal (the string does not move in the horizontal direction), but the net vertical force will lead to acceleration. The vertical components $F_v(x)$ and $F_v(x + \Delta x)$ can be found from the vector triangles shown in Fig. 4.2a and b as follows:

$$F_v(x) = -F_h \tan \alpha = -F_h y_x(x) \qquad (4.5)$$

and

$$F_v(x + \Delta x) = F_h y_x(x + \Delta x)$$

Denoting the mass per unit length of the string by ρ, the mass of the elementary string between x and $x + \Delta x$ is $\rho \, \Delta x$. The acceleration is

$$a = y_{tt}, \qquad (4.6)$$

the second temporal derivative of the displacement. Neglecting the weight of the string (i.e. disregarding gravity) the equation of motion (force = mass × acceleration) comes to

$$F_h[y_x(x + \Delta x) - y_x(x)] = \rho \, \Delta x \, y_{tt}. \qquad (4.7)$$

But, from the usual definition of derivatives,

$$\frac{y_x(x + \Delta x) - y_x(x)}{\Delta x} \cong y_{xx}. \qquad (4.8)$$

Substituting eqn (4.8) into (4.7) and dividing both sides by Δx, we obtain the sought-for partial differential equation for the displacement of a vibrating string:

$$y_{xx} = \frac{\rho}{F_h} y_{tt}. \qquad (4.9)$$

Introducing the notation

$$c^2 = \frac{F_h}{\rho}, \qquad (4.10)$$

our final form is

$$y_{xx} = \frac{1}{c^2} y_{tt}, \qquad (4.11)$$

which is known as the wave equation.

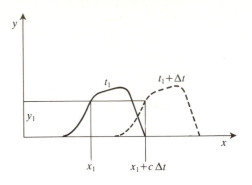

Fig. 4.3 A travelling wave profile

In order to show that this equation indeed has a wavelike solution, I shall just quote a simple solution,

$$y = f(x - ct), \qquad (4.12)$$

given in most textbooks concerned with waves, where f is an arbitrary function which depends on the argument $x - ct$.

Is this really a solution? Let's see. It is always easy to check whether we have got the solution of a particular differential equation. Let us differentiate the function twice, by x and t respectively:

$$\begin{aligned} y_x &= f'(x - ct), \qquad y_{xx} = f''(x - ct), \\ y_t &= -cf'(x - ct), \qquad y_{tt} = c^2 f''(x - ct), \end{aligned} \qquad (4.13)$$

where the prime means differentiation by the argument. Substituting the derivatives into eqn (4.11), we find that it is indeed satisfied.

What does $f(x - ct)$ mean? Taking some arbitrary function, $f(x - ct)$ is plotted in Fig. 4.3 as a function of x for a particular value of time, $t = t_1$. What will this function look like a time Δt later? In order to have the same value of y (say $y = y_{11}$) the value of x_1 must increase by $c\,\Delta t$, as shown by the dotted lines. The whole profile has moved a distance $c\,\Delta t$ in a time Δt. Hence c is the speed with which the whole profile moves. It is given by eqn (4.10). It is worth remembering that higher tension and lower mass density lead to higher speed.

The solution given by eqn (4.12) is valid for an infinitely long string. If we manage to produce a displacement profile somewhere on the line it will travel on the infinite line until the end of time. Thus, our solution of the wave equation represents a travelling wave of arbitrary profile.

We are now ready to look at more complicated problems. First of all let us consider a finite string which is fixed at $x = 0$ and $x = l$. The

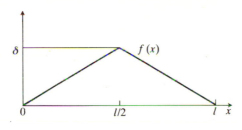

Fig. 4.4 The initial displacement

corresponding mathematical conditions are

$$y(0, t) = y(l, t) = 0, \qquad (4.14)$$

i.e. for all values of time the displacement at those points is zero.

Next we shall specify the initial velocity, i.e. the velocity at all values of x at $t = 0$. Assuming that the string is at rest at $t = 0$, the corresponding mathematical condition is

$$y_t(x, 0) = 0. \qquad (4.15)$$

If the initial displacement is also zero then the string will remain at rest for ever. If we want something to happen we need to specify an initial displacement. We shall assume the simple form shown in Fig. 4.4. It is the kind of form one may get by raising the middle of the string between one's fingers. Denoting the curve shown in Fig. 4.4 by $f(x)$, we can formally state the condition as

$$y(x, 0) = f(x). \qquad (4.16)$$

The problem has now been defined. We need to solve the partial differential equation (4.11) for $t > 0$ and $0 < x < l$ subject to the boundary condition (4.14) and the initial conditions (4.15) and (4.16).

Let us first solve the differential equation and worry about the boundary and initial conditions later. We shall attempt the solution by using the well-known and well-tried method of separation of variables. This is not the only method of course (we have already solved eqn (4.11) by guessing the solution in the form of eqn (4.12)), but it is the one most likely to lead to a result, and it is simple too. Accordingly, we shall assume the solution in the form

$$y(x, t) = X(x)T(t), \qquad (4.17)$$

whence

$$y_{xx} = TX'' \qquad (4.18)$$

and
$$y_{tt} = XT''. \tag{4.19}$$

Substituting eqns (4.18) and (4.19) into (4.11) and dividing by XT, we obtain

$$\frac{X''}{X} = \frac{1}{c^2}\frac{T''}{T}. \tag{4.20}$$

Now the left-hand side of the above equation is a function of x, and the right-hand side is a function of t only. They can only be equal to each other if both of them are equal to a constant, which we shall denote by v, i.e. we impose the condition that

$$\frac{X''}{X} = \frac{1}{c^2}\frac{T''}{T} = v. \tag{4.21}$$

Equation (4.21) may now be written as two separate ordinary differential equations:

$$X'' = vX \tag{4.22}$$

and

$$T'' = vc^2T. \tag{4.23}$$

Let us now commit ourselves concerning the sign of v. If v is negative we have the differential equations of simple harmonic motion with sinusoidal solutions. If v is positive the solutions are exponentials. I can tell you that the exponentials will not lead to a solution (one obvious difficulty with them is that the displacement of the string must be finite at all times, whereas exponentials might tend towards infinity) so let us take v negative, which we can ensure with the choice

$$v = -\mu^2 \tag{4.24}$$

where μ is a real constant.

In order to satisfy the boundary condition (4.14), it is obviously sufficient if it is satisfied by the space-varying function $X(x)$. The general solution is

$$X(x) = C_1 \cos \mu x + C_2 \sin \mu x. \tag{4.25}$$

At $x = 0$,

$$X(0) = C_1. \tag{4.26}$$

Thus, the condition $X(0) = 0$ can only be satisfied if $C_1 = 0$. In order to satisfy the condition $X(l) = 0$, we must have

$$X(l) = 0 = C_2 \sin \mu l. \tag{4.27}$$

Note that the $C_2 = 0$ solution is not available (because that would lead to the unexciting solution $X(x) = 0$); therefore, the only possibility is

$$\sin \mu l = 0, \tag{4.28}$$

which is satisfied whenever

$$\mu l = k\pi \quad \text{for} \quad k = 1,2,3. \tag{4.29}$$

Hence,

$$X(x) = \sin \frac{k\pi x}{l} \tag{4.30}$$

is a solution that satisfies the boundary conditions for any integer value of k.

Next let us try to satisfy the condition (4.15) by imposing conditions on the $T(t)$ function. The general solution of the differential equation (4.23) is

$$T(t) = A \cos c\mu t + B \sin c\mu t. \tag{4.31}$$

The condition (4.15) leads then to

$$T'(0) = 0 = c\mu B, \tag{4.32}$$

i.e. $B = 0$. Hence, a solution for the displacement which satisfies the boundary conditions and one of the initial conditions is

$$y(x, t) = X(x)T(t) = \sin \frac{k\pi x}{l} \cos \frac{ck\pi t}{l}. \tag{4.33}$$

Since our differential equation is linear, any sum of these functions is also a solution. Consequently, with arbitrary constants b_k the general solution for the displacement is

$$y(x, t) = \sum_{k=1}^{\infty} b_k \sin \frac{k\pi x}{l} \cos \frac{ck\pi t}{l}. \tag{4.34}$$

We still have one more condition to satisfy, namely the displacement at $t = 0$ is given by $f(x)$ shown in Fig. 4.4. Putting $t = 0$ into eqn (4.34) we end up with the equation

$$y(x, 0) = f(x) = \sum_{k=1}^{\infty} b_k \sin \frac{k\pi x}{l}. \tag{4.35}$$

You must admit now that eqn (4.35) looks like a Fourier series. The Fourier series of what? We haven't even got a periodic function on the left-hand side of eqn (4.35). True, but we can construct a periodic function if we think a little about it. The displacement is specified in the

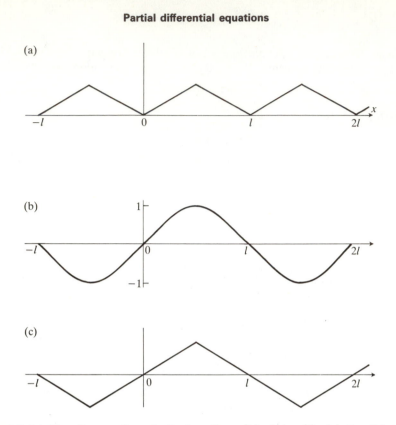

Fig. 4.5 (a) The "wrong" periodic function; (b) $\sin(\pi x/l)$; (c) the "right" periodic function

interval $0 < x < l$ and at $t = 0$ we know that it looks as shown in Fig. 4.4. It is easy to make a periodic function out of it. Just repeat the function outside the $(0, l)$ interval, as shown in Fig. 4.5a. Is this O.K.? I hope you will immediately realize that something is wrong. The function given in Fig. 4.5a is an even function, whereas the series in eqn (4.35) clearly represents an odd function. So what does the *right* periodic function look like? It must be odd and it must have a period of $2l$, because only then can the $k = 1$ choice give the fundamental component. This will be clear if you look at Fig. 4.5b where $\sin(\pi x/l)$ is plotted. Its period is $2l$. Now it is clear how the periodic function should be constructed. It must be half-range and odd, i.e. it will take the form shown in Fig. 4.5c. The Fourier series of this function may be found as

$$f(x) = \frac{8\delta}{\pi^2} \sum_{i=1}^{\infty} \frac{(-1)^{i-1}}{(2i-1)^2} \sin \frac{(2i-1)\pi x}{l}. \qquad (4.36)$$

Comparing eqns (4.35) and (4.36) it is clear that

$$b_{2i} = 0, \qquad b_{2i-1} = \frac{8\delta}{\pi^2} \frac{(-1)^{i-1}}{(2i-1)^2}. \tag{4.37}$$

Substituting b_k into eqn (4.34) we have our solution,

$$y(x, 0) = f(x) = \frac{8\delta}{\pi^2} \sum_{i=1}^{\infty} \frac{(-1)^{i-1}}{(2i-1)^2} \sin \frac{(2i-1)\pi x}{l} \cos \frac{c(2i-1)\pi t}{l}, \tag{4.38}$$

which satisfies all the boundary and initial conditions.

Having got the solution for the displacement it may be worth our while to plot it as a function of x with t as a parameter. For each value of t we have a Fourier series with coefficients

$$\frac{8\delta}{\pi^2} \frac{(-1)^{i-1}}{(2i-1)^2} \cos \frac{(2i-1)\pi ct}{l}. \tag{4.39}$$

What can we deduce from eqn (4.38)? We may notice that $\cos[c(2i - 1)\pi t/l]$ is a periodic function of t. Introducing normalized time through the relationship

$$t_N = \frac{ct}{l}, \tag{4.40}$$

the argument of the cosine function is $(2i - 1)\pi t_N$, i.e. the period is 2. Thus after every 2 units of normalized time we are back in the same situation. It is therefore sufficient to determine the displacement of the vibrating string in the interval $0 < t_N < 2$ and then everything repeats itself. In fact, it is sufficient to do the calculation for just one-half of the period, because by advancing one unit of normalized time the displacement function remains the same apart from a negative sign. This is because

$$\cos[(2i - 1)\pi(t_N + 1)] = -\cos(2i - 1)\pi t_N, \tag{4.41}$$

i.e. each term in the Fourier series will have its sign reversed.

We shall now evaluate the Fourier series given by eqn (4.38). In the past such a calculation required tremendous effort and that is the reason you find so few series evaluated in text books. Nowadays it can be done easily; one does not even need a fast computer, but old habits die hard. The emphasis in most text books is still on principles (not to mention lemmas) and not on detailed representation of the results. In my experience, applied scientists forget lemmas as soon as they put the book down: they will certainly remember the principles, but it is the examples to which they will turn when they need to solve a slightly (or perhaps considerably) different problem. So this is what we are going to do. We

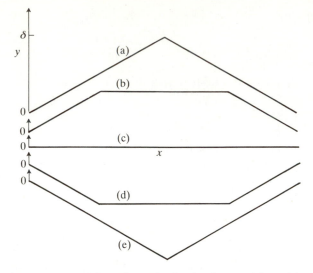

Fig. 4.6 Displacement as a function of x in the interval 0 to l for (a) $t_N = 0$, (b) $t_N = 0.25$, (c) $t_N = 0.5$, (d) $t_N = 0.75$, (e) $t_N = 1$

shall show lots of curves, each one representing the summation of the Fourier series given by eqn (4.38).

The displacement y for five discrete values of normalized time ($t_N = 0$, 0.25, 0.5, 0.75 and 1) is plotted in Figs. 4.6a–e. Is this what you expected? Well, you should have certainly expected that after a half period the displacement should change from that of Fig. 4.6a to its reverse shown in Fig. 4.6e. It follows from the mathematics as discussed above and it follows from the physics as well. In the absence of losses there is no mechanism to damp the amplitude of vibration; hence, it may be expected to swing out in the negative direction as much as its initial displacement in the positive direction. Furthermore, the curves are symmetric with respect to $x = l/2$, and that is expected because the initial excitation is symmetric too.

The only thing that might be unexpected is the sharp angular appearance of the curves, that they are always made up of straight lines. It is all right to have the straight lines at the beginning. One can bring a taut string to the position shown in Fig. 4.4, which accords with everyday experience, but once the string starts to vibrate those sharp corners should surely disappear. The reason for the discrepancy must be that in real life there are losses, whereas our model contains no losses. So we should remedy the situation and add losses to our model. But before we do that, let us see whether the angular feature will still prevail when the initial displacement is not symmetric.

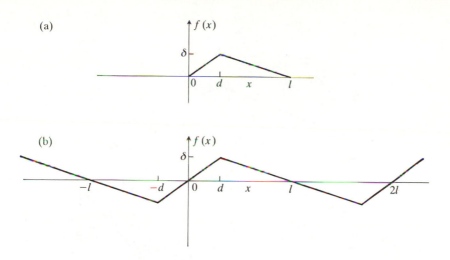

Fig. 4.7 (a) The initial displacement as a function of x. (b) The corresponding odd, half-range periodic function

We shall now choose the initial displacement in the form shown in Fig. 4.7a. The function is defined as

$$f(x) = \begin{cases} \delta x/4 & 0 < x < d \\ \delta \dfrac{l-x}{l-d} & \text{for} \quad d < x < l. \end{cases} \tag{4.42}$$

From what we have said before, the function that needs to be expanded into a Fourier series is the periodic function shown in Fig. 4.7b. After investing a moderate amount of mathematical effort, the Fourier coefficients may be obtained in the form (we have, again, an odd function and therefore the cosine components vanish)

$$b_k = \frac{2\delta l^2}{d(l-d)} \frac{1}{\pi^2 k^2} \sin \frac{k\pi d}{l}. \tag{4.43}$$

The complete solution for this initial value problem takes the following form

$$y(x, t_N) = \frac{2\delta l^2}{\pi^2 d(l-d)} \sum_{k=1}^{\infty} \frac{1}{k^2} \sin \frac{k\pi d}{l} \sin \frac{k\pi x}{l} \cos k\pi t_N. \tag{4.44}$$

It is the same solution as given by eqn (4.34) but now with b_k as given by eqn (4.43), and notice that we have changed to normalized time. The time variation may be seen still to be periodic with period 2.

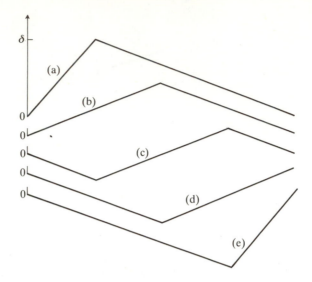

Fig. 4.8 Displacement as a function of x in the interval 0 to l for (a) $t_N = 0$, (b) $t_N = 0.25$, (c) $t_N = 0.5$, (d) $t_N = 0.75$, (e) $t_N = 1$

The numerical solutions for $d/l = 0.25$ and $t_N = 0$, 0.25, 0.5, 0.75, 1 are shown in Figs. 4.8a–e. Note that the solutions are still made up of straight lines and that a half period later the same asymmetry asserts itself but in the opposite direction.

Losses

It has been suggested that lack of losses might be responsible for the sharp corners, so it is worth seeing what happens when we introduce losses. How can we do that? Normally, losses are introduced in a phenomenological manner by adding to the differential equation a term proportional to the first temporal derivative. We shall do just that and change our partial differential equation (4.11) to the form

$$y_{xx} = \frac{1}{c^2}(y_{tt} + 2by_t),$$ (4.45)

where b is now the parameter responsible for losses.

We may again find the solution by the method of separating the variables. Provided we have the same boundary conditions (given by eqn (4.14)), the solution for $X(x)$ remains the same as given by eqn (4.30). The differential equation for $T(t)$ will, however, be different, because the first derivative term will make its appearance. It takes the form

$$T'' + 2bT' + \mu^2 c^2 T = 0, \tag{4.46}$$

where, as before,

$$\mu = \frac{k\pi}{l}. \tag{4.47}$$

The solution of the above differential equation with initial conditions

$$T(0) = 1 \quad \text{and} \quad T'(0) = 0 \tag{4.48}$$

may be obtained as

$$T(t) = \exp(-gt_N)\left(\cos \gamma t_N + \frac{g}{\gamma}\sin \gamma t_N\right), \tag{4.49}$$

where

$$\gamma = (k^2\pi^2 - g^2)^{1/2}, \qquad g = \frac{bl}{c}. \tag{4.50}$$

The complete solution for the initial displacement of Fig. 4.7a may now be written as

$$y(x, t) = \frac{2\delta^2 l \exp(-gt_N)}{\pi^2 d(l-d)} \sum_{k=1}^{\infty} \frac{1}{k^2}\sin\frac{\pi kd}{l}\sin\frac{\pi kx}{l}\left(\cos \gamma t_N + \frac{g}{\gamma}\sin \gamma t_N\right). \tag{4.51}$$

If we evaluate eqn (4.51) for small losses, we find (just believe me, there is no point in plotting the curves) that only the term $\exp(-gt_N)$ has significant influence. In other words, as time goes on all the amplitudes of the vibration gradually become smaller but there is hardly any change in the shape of the displacement curves. They are still composed of straight lines.

So losses on their own will not make the displacement curves more 'natural' looking. Does this mean that we are on the wrong track? No, but we need to introduce one more modification into our model. Obviously, the way to get rid of sharp corners is to reduce the relative amounts of higher harmonics. We can do that by assigning higher losses to the higher harmonics, equivalent to the kind of filter we talked about in Section 2.11. Assuming that the loss is simply proportional to frequency,† we shall use the form

$$g = g_0 k. \tag{4.52}$$

---

† We shall not dwell on this assumption here, but we shall, for transmission lines in Section 4.4, use a more realistic formula for the frequency dependence of loss.

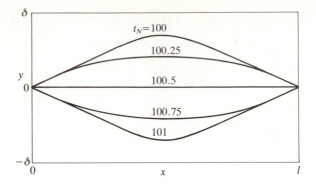

Fig. 4.9 Displacement as a function of x for the lossy case ($g_0 = 0.0025$) for the half period starting at $t_N = 100$ for the symmetric initial displacement of Fig. 4.4

Replacing g in eqn (4.51) by its k-dependent form of eqn (4.52) and bringing the $\exp(-gt_N)$ term within the summation sign, we may now plot $y(x, t)$. It is shown in Fig. 4.9 for $g = 0.0025$ and for the symmetric case ($d/l = 0.5$) at $t_N = 100$, 100.25, 100.5, 100.75 and 101. Comparing Fig. 4.9 with Fig. 4.6 it may be clearly seen that after 50 periods the amplitude of vibration has declined and, in addition, the sharp corners have disappeared. Advancing a further 100 periods the displacement is plotted in Fig. 4.10 at $t_N = 300$, 300.25, 300.5, 300.75 and 301. This now looks a pure sinusoidal. Is it? We can easily find out by taking the ratio of the

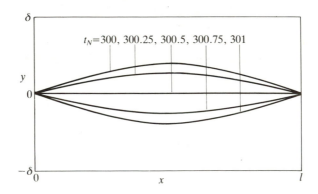

Fig. 4.10 Displacement as a function of x for the lossy case ($g_0 = 0.0025$) for the half period starting at $t_N = 300$ for the symmetric initial displacement of Fig. 4.4

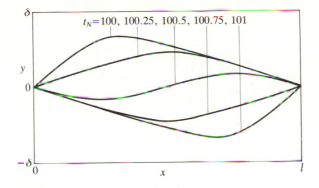

Fig. 4.11 Displacement as a function of x for the lossy case ($g_0 = 0.0025$) for the half period startig at $t_N = 100$ for the asymmetric initial displacement of Fig. 4.7a

first non-zero harmonic to the fundamental. In the present symmetric case the second harmonic is absent, therefore the relevant measure is

$$\frac{\text{Third harmonic}}{\text{Fundamental}} = \frac{\exp(-3 \times 0.0025 \times 300) \times (1/9)}{\exp(-0{\cdot}0025 \times 300)} = 0.024, \quad (4.53)$$

which indeed shows that the third harmonic makes very little contribution.

We may conclude that when losses are higher for higher frequencies, the sharp corners soon disappear, and if we wait long enough (about 150 periods in the present example), only the fundamental component will survive.

Are these conclusions valid for the asymmetric excitation as well? Let us see. We shall take $d/l = 0.25$, the same loss parameter, $g_0 = 0.0025$, and again plot the displacement in Fig. 4.11 for five discrete values within the half period starting at $t_N = 100$. It may be seen quite clearly that the main features seen in Fig. 4.8 are still present but the amplitudes are smaller and the sharp corners have been rounded off. At $t_N = 300$ (Fig. 4.12) and at $t_N = 600$ (Fig. 4.13, note that the vertical scale is expanded by a factor of 2), the decay of the amplitudes continues and the asymmetry becomes less. If we go as far as $t_N = 1200$ (Fig. 4.14), we seem to be left with the fundamental component only. To ascertain this let us compare again the amplitude of the first non-zero harmonic (second in the present case) with the fundamental. It may be worked out to give

$$\frac{\text{Second harmonic}}{\text{Fundamental}} \cong \frac{\exp(-2 \times 0.0025 \times 1200) \times \frac{1}{4}}{\sin(\pi/4) \times \exp(-0.0025 \times 1200)} = 0.017. \quad (4.54)$$

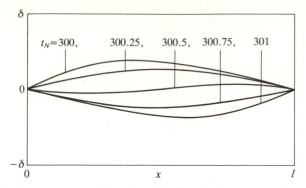

Fig. 4.12 Displacement as a function of x for the lossy case ($g_0 = 0.0025$) for the half period starting at $t_N = 300$ for the asymmetrical initial displacement of Fig 4.7a

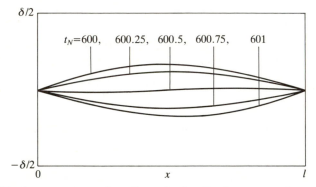

Fig. 4.13 Displacement as a function of x for the lossy case ($g_0 = 0.0025$) for the half period starting at $t_N = 600$ for the asymmetrical initial displacement of Fig. 4.7a

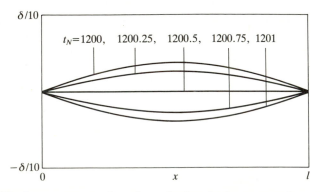

Fig. 4.14 Displacement as a function of x for the lossy case ($g_0 = 0.0025$) for the half period starting at $t_N = 1200$ for the asymmetrical initial displacement of Fig. 4.7a

(a)

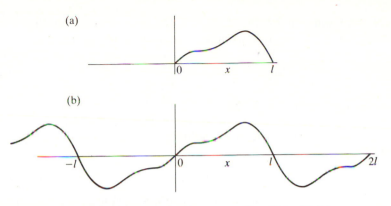

(b)

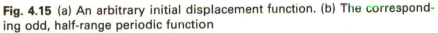

Fig. 4.15 (a) An arbitrary initial displacement function. (b) The corresponding odd, half-range periodic function

We may thus conclude again that, in spite of the asymmetric excitation, if we wait long enough only the fundamental component will survive.

Further generalizations

We have so far solved two examples for slightly different initial conditions. It is easy to see, however, that these two examples were not in any way distinguished. The same technique of solution could be applied whatever the chosen displacement $f(x)$ at $t = 0$. For example, if $f(x)$ is of the form shown in Fig. 4.15a then the relevant coefficients can be obtained by expanding the function shown in Fig. 4.15b (the corresponding odd, half-range periodic function) into a Fourier series. The function being odd, the cosine terms vanish and we may obtain the coefficients of the sine series from eqn (2.87) as follows:

$$b_k = \frac{2}{2l} \int_0^{2l} f(x) \sin \frac{2\pi kx}{2l} \, dx$$

$$= \frac{2}{l} \int_0^{l} f(x) \sin \frac{k\pi x}{l} \, dx. \tag{4.55}$$

To have all the relevant conditions together, I shall repeat here the boundary conditions

$$y(0, t) = y(l, t) = 0 \tag{4.56}$$

we imposed earlier, and the conditions for the initial displacement and initial velocity

$$y(x, 0) = f(x), \qquad y_t(x, 0) = 0. \tag{4.57}$$

The general solution for the displacement of a lossless vibrating string satisfying the above conditions may be written in the form

$$y(x, t) = \sum_{k=1}^{\infty} b_k \sin\frac{k\pi x}{l} \cos\frac{k\pi ct}{l}, \tag{4.58}$$

where b_k is given by eqn (4.55).

Finally, we shall consider the case when all the conditions given above are the same apart from the initial velocity, which is taken as an arbitrary function

$$y_t(x, 0) = h(x). \tag{4.59}$$

To solve this new problem we may proceed again in the old familiar manner. We may find $X(x)$ in the form of eqn (4.30) and $T(t)$ in the form of eqn (4.31), which we reproduce below:

$$T(t) = A \cos \mu ct + B \sin \mu ct. \tag{4.60}$$

The difference in comparison with the previous cases is that we now have two initial conditions and therefore both sets of coefficients A_k and B_k will be necessary for the solution.

The form for the general solution is

$$y(x, t) = \sum_{k=1}^{\infty} \sin\frac{k\pi x}{l} \left(A_k \cos\frac{k\pi ct}{l} + B_k \sin\frac{k\pi ct}{l} \right). \tag{4.61}$$

Let us first satisfy the condition for the initial displacement at $t = 0$. From eqn (4.61) we obtain the relationship

$$y(x, 0) = f(x) = \sum_{k=1}^{\infty} A_k \sin\frac{k\pi x}{l}, \tag{4.62}$$

whence (following the recipe used earlier)

$$A_k = \frac{2}{l} \int_0^l f(x) \sin\frac{k\pi x}{l}\, dx. \tag{4.63}$$

To impose the condition for the initial velocity we need to find the velocity by differentiating eqn (4.61). We obtain

$$y_t(x, t) = \sum_{k=1}^{\infty} \left(\sin\frac{k\pi x}{l} \right) \frac{ck\pi}{l} \left(-A_k \sin\frac{k\pi ct}{l} + B_k \cos\frac{k\pi ct}{l} \right). \tag{4.64}$$

Should we exercise some care here? Can eqn (4.61) always be differentiated? Well, as the displacement of a string describes a physical problem, it cannot have any horrible discontinuities. There is no need to worry.

Returning now to the initial velocity, we may obtain an expression for it by substituting $t = 0$ into eqn (4.64), yielding

$$y_t(x, 0) = h(x) = \sum_{k=1}^{\infty} \frac{ck\pi}{l} B_k \sin \frac{k\pi x}{l}. \tag{4.65}$$

Now, it is $h(x)$ for which an odd half-range periodic function needs to be constructed. The technique is the same as before, leading to

$$\frac{ck\pi}{l} B_k = \frac{2}{l} \int_0^l h(x) \sin \frac{k\pi x}{l} \, dx, \tag{4.66}$$

whence B_k can be determined.

   Owing to lack of space I shall not be able to give here numerical examples, but if you want to follow what happens when the initial velocity is specified, attempt Exercises 4.3 and 4.4.

4.3   Heat conduction

We shall again start by deriving the relevant partial differential equation for heat conduction. It will be based on two empirical observations, namely (i) the change of heat stored in an element is proportional to the change in temperature, and (ii) the amount of heat crossing a surface per unit time is proportional to the temperature gradient.

   For simplicity, we shall look at one-dimensional problems only, in which there is variation only in the x direction. For deriving the equations we shall take a slab of thickness Δx, as shown in Fig. 4.16. The heat entering the slab in a time Δt is

$$-Ku_x(x, t) \, \Delta t, \tag{4.67}$$

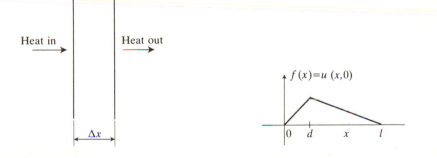

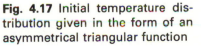

Fig. 4.16 Schematic representation of heat flow for a slab of
elementary thickness

Fig. 4.17 Initial temperature distribution given in the form of an
asymmetrical triangular function

where u is the temperature (u_x is the temperature gradient) and K is a proportionality constant. The negative sign comes from the following considerations. Positive temperature gradient means that the temperature increases in the positive x direction. Since heat may be expected to flow from higher towards lower temperatures, a positive gradient indicates a heat flow in the negative x direction. Therefore a negative sign must be attached if heat flows into the slab of Fig. 4.16. Similarly, heat leaving the slab in a time Δt is

$$-Ku_x(x + \Delta x, t)\, \Delta t. \tag{4.68}$$

Consequently, the net heat leaving the slab is

$$-K[u_x(x + \Delta x, t) - u_x(x, t)]\, \Delta t. \tag{4.69}$$

The change in the heat stored in the slab is equal to

$$C\, \Delta u\, \Delta x, \tag{4.70}$$

where C is another proportionality constant.

The change in the heat stored must be equal to the negative of the net heat leaving, leading to

$$C\, \Delta u\, \Delta x = K[u_x(x + \Delta x, t) - u_x(x, t)]\, \Delta t \tag{4.71}$$

which, after rearrangement, takes the form

$$\frac{\Delta u}{\Delta t} = \frac{K}{C} \frac{u_x(x + \Delta x, t) - u_x(x, t)}{\Delta x}. \tag{4.72}$$

Introducing the notation $a^2 = K/C$, we obtain in the limit of $\Delta x \to 0$ and $\Delta t \to 0$,

$$u_t = a^2 u_{xx}, \tag{4.73}$$

known as the one-dimensional heat conduction equation.

As our first example we shall choose a conducting rod of length l whose sides are insulated so that no heat conduction takes place through the side surfaces. This is a good approximation to a one-dimensional problem. As for boundary conditions, we shall keep the two ends of the rod at zero temperature, yielding the mathematical conditions

$$u(0) = u(l) = 0. \tag{4.74}$$

At $t = 0$ we shall assume that the temperature distribution is given by the asymmetrical triangular function shown in Fig. 4.17, defined mathematically by eqn (4.42). It may be noticed that we take the same boundary and initial conditions as in one of our examples of the vibrating string, with one exception because in that problem we also specified the initial velocity, i.e. the temporal derivative at $t = 0$. In the present example we cannot specify the time derivative; it will come out of the calculations. This is because in the differential equation for heat conduction the time derivative is only of first order, we can impose only one condition, and that will be the initial distribution of temperature.

We shall again start by separating the variables, i.e. by assuming the solution in the form

$$u(x, t) = X(x)T(t). \tag{4.75}$$

Substituting eqn (4.78) into (4.73) and dividing both sides by a^2XT we obtain

$$\frac{1}{a^2}\frac{T'}{T} = \frac{X''}{X}. \tag{4.76}$$

Again using the argument that for eqn (4.76) to be valid both sides must be equal to a constant, and the sign of the constant is determined by the requirement that the temporal function cannot grow with increasing t, eqn (4.76) can be turned into the ordinary differential equations

$$X'' + \mu^2 X = 0 \tag{4.77}$$

and

$$T' + \mu^2 a^2 T = 0. \tag{4.78}$$

We have met eqn (4.77) before. Both the differential equation and the boundary conditions (at $x = 0$ and $x = l$) are the same as in our example for the vbirating string, hence the general solution is the same as that given by eqn (4.30):

$$X = \sin \mu x, \qquad \mu = \frac{k\pi}{l}, \tag{4.79}$$

where k can be any positive integer.

We can now turn our attention to eqn (4.78), the differential equation in t which is now of first order. The solution is

$$T = \exp\left(-\frac{k^2\pi^2 a^2 t}{l^2}\right). \tag{4.80}$$

Hence, the general solution of eqn (4.73) satisfying the boundary conditions is

$$u(x, t) = \sum_{k=1}^{\infty} b_k \exp\left(-\frac{k^2\pi^2 a^2 t}{l^2}\right) \sin \frac{k\pi x}{l}, \tag{4.81}$$

where b_k are unknown coefficients to be determined.

For $t = 0$ the initial temperature distribution is given by $f(x)$ of Fig.

4.17, leading to the equation

$$f(x) = \sum_{k=1}^{\infty} b_k \sin \frac{k\pi x}{l}. \tag{4.82}$$

The mathematical problem is now identical with the one we had in the previous section. The coefficients are given by eqn (4.44), leading to the solution in the form

$$u(x, t_N) = \frac{2\delta l^2}{\pi^2 d(l-d)} \sum_{k=1}^{\infty} \frac{1}{k^2} \sin \frac{k\pi d}{l} \sin \frac{k\pi x}{l} \exp(-k^2 t_N), \tag{4.83}$$

where we have introduced a normalized time through the relationship

$$t_N = \left(\frac{a\pi}{l}\right)^2 t. \tag{4.84}$$

We have now got the solution. How will the temperature distribution vary as a function of time? It may be seen from eqn (4.83) by inspection that the temperature tends to zero as $t_N \rightarrow \infty$. How will its shape vary during decay? After the lapse of a sufficiently small time the temperature distribution should of course be similar to the initial distribution, $f(x)$. But the higher harmonics will decay much faster than the fundamental, as indicated by the k^2 term in the exponential in eqn (4.83). Hence, we may expect that the asymmetry will quickly disappear and only the fundamental component will remain.

Let us see now whether our expectations are borne out by the numerical calculations. Equation (4.83) is plotted for $t_N = 0$, 0.1, 0.4, 1 and 2 in Fig. 4.18a, and for $t_N = 2$, 2.5 and 3 in Fig. 4.18b on a vertical scale expanded by a factor 12. It may indeed be seen that the decay is fast and particularly that the asymmetry quickly vanishes. At $t_N = 1$ the temperature distribution already looks sinusoidal. To find a better measure we can work out the ratio of the second harmonic component to the fundamental component. It comes to 0.07 (still noticeable) at $t_N = 0.4$, to 0.0065 at $t_N = 1$, and to 0.00019 at $t_N = 2$.

We may conclude, unsurprisingly, that there are no oscillations when only a first-order time derivative is present in the differential equation. The decay is fast, and the decay of the higher harmonics is particularly fast.

In our next example we shall look again at one dimensional heat conduction, but the time variation will be imposed at one particular point. We shall take a slab $0 < x < l$ and assume that on the surface (at $x = 0$) the temperature varies as a function of time in the manner shown in Fig. 4.19 (in practice this may be approximately realized by illuminating the surface of a heat-conducting and heat-absorbing medium by

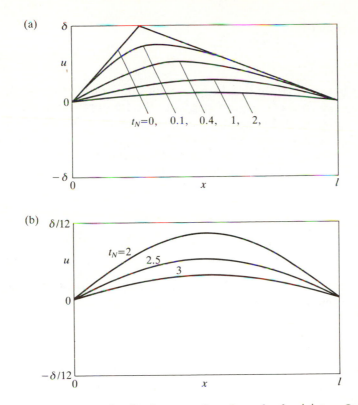

Fig. 4.18 Temperature distribution as a function of x for (a) $t_N = 0$, 0.1, 0.4, 1, 2; (b) $t_N = 2$, 2.5, 3—note that the scale has been expanded by a factor of 12

pulsed laser light). There are no initial conditions now; $t = 0$ is in no way distinguished. We are concerned with the steady-state solution, i.e. what will the temperature distribution be as a function of space and time inside the medium in response to the surface excitation given in Fig. 4.19.

Again, we need to solve eqn (4.73). By separation of variables (i.e. assuming the solution in the form of eqn (4.75)) and dividing both sides

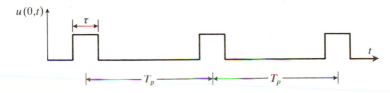

Fig. 4.19 Temperature variation as a function of time on the surface, $x = 0$

by XT we obtain

$$\frac{T'}{T} = a^2 \frac{X''}{X}. \tag{4.85}$$

We know by now that we should make both sides equal to a constant, but how should we choose the constant? Since the temporal variation at $x = 0$ will be constructed by adding all the harmonics of the frequency $\omega = 2\pi/T_p$ (we have here introduced the notation T_p for the period in order to distinguish it from the temporal function T), the logical choice for the constant is $jk\omega t$. The solution of the differential equation

$$\frac{T'}{T} = jk\omega \tag{4.86}$$

is then

$$T = C_k \exp jk\omega t, \tag{4.87}$$

where C_k is a constant. The differential equation in the spatial variable may be written in the form

$$X'' - \frac{jk\omega}{a^2} X = 0, \tag{4.88}$$

which is again the differential equation of simple harmonic motion, although the constant is now imaginary which actually makes a lot of difference.

Formally, we may assume the solution in the form

$$X = \exp \lambda x \tag{4.89}$$

which, substituted into eqn (4.88), leads to the algebraic equation

$$\lambda^2 = \frac{jk\omega}{a^2}. \tag{4.90}$$

For $k > 0$ the square root is taken as

$$\lambda = \pm \frac{1+j}{a} \sqrt{\frac{\omega |k|}{2}}. \tag{4.91}$$

Since an exponentially growing solution is physically impossible, we can take only the negative sign in eqn (4.91).

For $k < 0$ the square root in eqn (4.90) is taken as

$$\lambda = \pm \frac{1-j}{a} \sqrt{\frac{\omega |k|}{2}}, \tag{4.92}$$

of which, again, only the negative sign has physical meaning.

The complete solution, summing over all values of k, is

$$u(x, t) = \sum_{k=-\infty}^{\infty} C_k \, e^{\lambda x} \exp jk\omega t. \tag{4.93}$$

At $x = 0$ the temporal variation of temperature should be equal to $f(t)$ of Fig. 4.19, yielding the relationship

$$f(t) = \sum_{k=-\infty}^{\infty} C_k \exp jk\omega t. \tag{4.94}$$

We worked out the Fourier series of $f(t)$ in Section 2.9. It is

$$f(t) = \frac{1}{\pi} \sum_{k=-\infty}^{\infty} \frac{1}{k} \sin \frac{\pi k \tau}{T_p} \exp jk\omega t. \tag{4.95}$$

Comparing eqn (4.94) with eqn (4.95), C_k may be determined. It may be seen that C_k is independent of the sign of k. It should, however, be remembered that λ has different forms (eqn (4.91) for positive k, and eqn (4.92) for negative k) depending on the sign of k. It is therefore convenient to divide the summation in eqn (4.93) into two parts, one sum for negative k, and another sum for positive k, as shown below:

$$u(x, t) = \sum_{k=-\infty}^{-1} C_k \exp\left[-(1-j) \sqrt{\frac{\omega |k|}{2a^2}} x \right] \exp jk\omega t$$

$$+ \sum_{k=1}^{\infty} C_k \exp\left[-(1+j) \sqrt{\frac{\omega k}{2a^2}} x \right] \exp jk\omega t, \tag{4.96}$$

where the $k = 0$ term, not having physical significance, has been disregarded. Replacing k by $-k$ in the first summation of eqn (4.96), we shall end up with one summation only, which takes the form

$$u(x, t) = \sum_{k=1}^{\infty} C_k \exp\left(-\sqrt{\frac{k\omega}{2a^2}} x \right) \left\{ \exp jk\left(\sqrt{\frac{\omega}{2a^2 k}} x - \omega t \right) \right.$$

$$+ \exp\left[-jk\left(\sqrt{\frac{\omega}{2a^2 k}} x - \omega t \right) \right] \right\} = \frac{2}{\pi} \sum_{k=1}^{\infty} \frac{1}{k} \sin \frac{\pi k \tau}{T_p}$$

$$\times \exp\left(-\sqrt{\frac{k\omega}{2a^2}} x \right) \cos\left[k\omega \left(t - \frac{x}{\sqrt{2a^2 \omega k}} \right) \right]. \tag{4.97}$$

It may be seen from the argument of the cosine function that we have a travelling wave (this is indicated by the form $t - x/v$, where v may be regarded as the velocity of the wave) and the exponent shows that the wave is also attenuated. The attenuation depends on k: the higher the harmonic, the larger is the attenuation. Hence, far enough into the material only the fundamental component will remain. If you are interested to see how the wave form changes as the heat wave propagates into the interior of the material, you can plot eqn (4.97). I shall not do the numerical calculations here because I intend to go into more detail concerning a similar, and in practice much more important, problem, namely the attenuation and propagation of an electric wave on a transmission line.

4.4 Transmission lines

Transmission lines usually consist of two conductors (an example is the so-called two-wire or Lecher line, another example is a coaxial cable) which carry electricity, i.e. certain currents flow in them and there are certain voltages between them. These currents and voltages vary with time t and depend of course on the spatial coordinate x. In general, we wish to know the values of the currents and voltages at a given time t, at a given point x. Thus, our independent variables are x and t, and the dependent variables are $I(x, t)$ and $V(x, t)$ where I and V again denote current and voltage.

How will I and V be related to each other? It will depend on the parameters of the line, namely on the inductance L, the capacitance C, the resistance R, and the conductance G, all of them per unit distance. Is it obvious that these are the right parameters? Not at first sight. It might be worthwhile to dwell on these parameters a little longer.

It is probably easiest to account for the resistance R. In all conductors (even in superconductors at high frequencies) there is a resistance to the progress of current. The voltage produced in an elementary line of length dx is given by Ohm's law, $V = IR\,dx$. WIth the conductance G a little care needs to be exercised. It is not the reciprocal of R; it is a conductance in its own right, often called a leakage conductance. It occurs because the medium between the conductors is not a perfect insulator. Some current will leak through that medium. At a voltage V, the ensuing current for an elementary line of length dx is equal to $I = VG\,dx$.

Remember that currents generate magnetic fields and that changing magnetic fields induce voltages. The relationship between current and voltage is

$$V = L\,dx\,\frac{\partial I}{\partial t}, \tag{4.98}$$

where $L\,dx$ is the inductance for an elementary line of length dx. Note that in eqn (4.98) we have used partial differentiation because t is not the only variable.

Owing to the capacitance, voltage and charge are also related to each

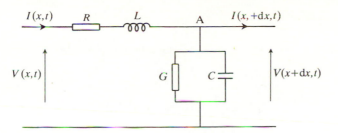

Fig. 4.20 Equivalent circuit for a transmission line of elementary length

other. For an elementary line of length dx the charge Q is equal to $Q = VC\,dx$. Since current is the temporal rate of change of charge, our equation is

$$I = \frac{\partial Q}{\partial t} = C\,dx\,\frac{\partial V}{\partial t}. \tag{4.99}$$

Having got our four relationships between current and voltage we still need to find a formal representation of an element of the transmission line. It is taken in the form shown in Fig. 4.20 where the elementary line is taken between the points x and $x + dx$. The voltage and current at time t at the point x are $V(x, t)$ and $I(x, t)$ and at the point $x + dx$ they are $V(x + dx, t)$ and $I(x + dx, t)$ respectively. The arrangement of the circuit elements in Fig. 4.20 is fairly obvious: the inductance and resistance are in series, whereas the capacitance and the conductance are in parallel.

Let us now derive the corresponding equations. The voltage difference between the two ends is caused by the voltage drop across the resistance and the inductance, i.e.

$$V(x, t) - V(x + dx, t) = R\,dx\,I(x, t) + L\,dx\,\frac{\partial I(x, t)}{\partial t}. \tag{4.100}$$

Next we need to establish the current balance. According to Kirchhoff's law the current flowing into a junction must be equal to the current flowing out. Considering that at point A the current flowing towards the other conductor is

$$I_p = G\,dx\,V(x + dx, t) + C\,dx\,\frac{\partial V(x + dx, t)}{\partial t}, \tag{4.101}$$

our equation becomes

$$I(x, t) = G\,dx\,V(x + dx, t) + C\,dx\,\frac{\partial V(x + dx, t)}{\partial t} + I(x + dx, t). \tag{4.102}$$

But, for small dx,

$$V(x + dx, t) = V(x, t) + \frac{\partial V(x, t)}{\partial x} dx \qquad (4.103)$$

and

$$I(x + dx, t) = I(x, t) + \frac{\partial I(x, t)}{\partial x} dx. \qquad (4.104)$$

Substituting eqn (4.103) into eqn (4.100) we find

$$-\frac{\partial V(x, t)}{\partial x} dx = R \, dx \, I(x, t) + L \, dx \, \frac{\partial I(x, t)}{\partial t}. \qquad (4.105)$$

It may be seen now that each term is proportional to dx; hence, dx can be dropped and we end up with the partial differential equation

$$-\frac{\partial V}{\partial x} = RI + L \frac{\partial I}{\partial t}, \qquad (4.106)$$

whence we have omitted the brackets indicating that V and I are functions of x and t.

Next, let us devote our attention to eqn (4.102). In view of eqn (4.104) it takes the form

$$-\frac{\partial I(x, t)}{\partial x} dx = G \, dx \, V(x + dx, t) + C \, dx \, \frac{\partial V(x + dx, t)}{\partial t}. \qquad (4.107)$$

We could again use the form of $V(x + dx, t)$ as given by eqn (4.103), but it is not really necessary. We are interested only in terms first-order in dx and hence we may replace $V(x + dx, t)$ in eqn (4.107) (as it is to be multiplied by dx anyway) simply by $V(x, t)$, and the same applies to the derivative of $V(x + dx, t)$. Further omitting the brackets, eqn (4.107) reduces to

$$-\frac{\partial I}{\partial x} = GV + C \frac{\partial V}{\partial t}. \qquad (4.108)$$

The differential equations to solve are now eqns (4.106) and (4.108). In the usual analysis of transmission lines it is assumed at this stage that the excitation is at a single frequency ω and the solution of the differential equation gives β, the corresponding propagation coefficient. This solution is relevant even in the case when the input carrier frequency is modulated, because the modulating frequency is nearly always much less than the carrier frequency. However, the present course is concerned with the properties of Fourier series and therefore we shall abandon the usual practice and investigate the case when the modulation is in the form of so-called base band signals, i.e. when there is no carrier wave. The set of input signals may then be represented by a Fourier series in which the fundamental frequency is the pulse repetition frequency.

I do realize that in the preceding discussion there has been a lot of communications engineering jargon. It is not really important. What matters is that the input will be in the form of a Fourier series, and the logical thing is to assume the solution of the transmission line equations in the analogous form of a Fourier series of travelling waves,

$$V(x, t) = \sum_{k=-\infty}^{\infty} A_k \exp j(k\omega t - \beta_k x) \tag{4.109}$$

and

$$I(x, t) = \sum_{k=-\infty}^{\infty} B_k \exp j(k\omega t - \beta_k x), \tag{4.110}$$

where β_k is the unknown propagation coefficient at the frequency $k\omega$.

Substituting eqn (4.109) and (4.110) into eqns (4.106) and (4.108) we obtain, for each value of k,

$$-A_k(-j\beta_k) = (R + jk\omega L)B_k \tag{4.111a}$$

and

$$-B_k(-j\beta_k) = (G + jk\omega C)A_k. \tag{4.111b}$$

This is a set of linear algebraic equations in A_k and B_k which have a solution only if the determinant vanishes, i.e. if

$$\begin{vmatrix} j\beta_k & -(R + jk\omega L) \\ -(G + jk\omega C) & j\beta_k \end{vmatrix} = 0, \tag{4.112}$$

whence, solving for β_k, we obtain

$$\beta_k = \pm j[(R + jk\omega L)(G + jk\omega C)]^{1/2}. \tag{4.113}$$

The above square root can in general be evaluated to give a complex number for β_k. The real part is then related to propagation and the imaginary part to attenuation, i.e. to the decay of the wave. This may be seen immediately by rewriting eqn (4.109) in the form

$$V(x, t) = \sum_{k=-\infty}^{\infty} A_k \exp[-(\text{Im } \beta_k)x]\exp j[k\omega t - (\text{Re } \beta_k)x]. \tag{4.114}$$

Next, let us look at the initial conditions. At $x = 0$, at the input of the transmission line, we shall specify the voltage as a periodic function of time represented by its Fourier series

$$V(0, t) = \sum_{k=-\infty}^{\infty} c_k \exp j\omega t. \tag{4.115}$$

Comparing eqn (4.109) with (4.115) it may be seen that

$$A_k = c_k \tag{4.116}$$

and B_k may be determined from eqn (4.111) as

$$B_k = \frac{\mathrm{j}\beta_k}{R + \mathrm{j}k\omega L} A_k. \tag{4.117}$$

Is this the end of our derivation? Yes, it is; we have determined all the unknowns and so we can investigate a few examples.

1. *Lossless case.* $R = 0$ and $G = 0$, then

$$\beta_k = \omega k \sqrt{LC} \tag{4.118}$$

and the solution for the voltage is

$$V(x, t) = \sum_{k=-\infty}^{\infty} c_k \exp \mathrm{j}k\omega(t - x\sqrt{LC}), \tag{4.119}$$

whence we may deduce that the velocity of the travelling waves is

$$v = \frac{1}{\sqrt{LC}}. \tag{4.120}$$

Note that the velocity is independent of the frequency, therefore all harmonics propagate with the same velocity. The waveform does not distort as the wave propagates down the transmission line. In communications engineering jargon, one says that the line is dispersionless.

2. *A real transmission line.* For our next example we take a coaxial cable in which there is skin effect and which is filled with a real material. Consequently, both R and G become frequency-dependent. We shall not give here the details of the calculation. For our purpose it is sufficient to say that with good approximation $\mathrm{Re}\,\beta_k$ remains the same as given by eqn (4.118) and the imaginary part takes the form

$$\mathrm{Im}\,\beta_k = 0.933 \times 10^{-7}\sqrt{k\omega} + 0.25 \times 10^{-11}\,k\omega \quad \mathrm{m}^{-1}. \tag{4.121}$$

Since $\mathrm{Re}\,\beta_k$ is unchanged, it is still true that all Fourier components travel with the same velocity. However, $\mathrm{Im}\,\beta_k$ is frequency-dependent; therefore each harmonic will have a different amount of decay. Making the length l of our transmission line to be equal to 70 km, we find for the total attenuation

$$(\mathrm{Im}\,\beta_k)l = 6.53 \times 10^{-3}\sqrt{k\omega} + 1.75 \times 10^{-7}\,k\omega. \tag{4.122}$$

Taking $\omega = 6.28 \times 10^3$ radian s^{-1} ($f = 10^3$ Hz) this comes, for example, to

0.517 at $k = 1$, to 0.732 at $k = 2$, and to 0.895 at $k = 3$. This means that at the output the voltage at the fundamental component will have declined by a factor of $\exp(0.517) = 1.68$, the second harmonic by a factor of $\exp(0.732) = 2.08$, and the third harmonic by a factor of $\exp(0.895) = 2.45$. Obviously, the attenuation increases as k increases. You will realize that this is equivalent to a filter. Using the notation we had for describing filters, the relationship is now

$$F_k = \exp[-(\text{Im } \beta_k)l]. \tag{4.123}$$

The absolute amount of attenuation is of course very important. If the attenuation is too large it may not be possible to recover the signal from noise, but for our purpose the absolute attenuation is irrelevant. We are primarily interested in the distortion of the waveform. We shall therefore assume that the fundamental component at the output will be the same as at the input, but we shall take into account the difference in attenuation between the harmonics. Returning to the example just given, the second harmonic will decline relative to the fundamental by a factor of $\exp(0.732 - 0.517) = 1.24$, and the third harmonic by a factor of $\exp(0.895 - 0.517) = 1.46$.

For our input periodic function we shall choose the one shown in Fig. 1.7a (reproduced for easier comparison in Fig. 4.21a) in which the pulse is on for one-quarter of the time, and off for three-quarters of the time. We shall illustrate the distortion that arises by plotting the first 17 components of this waveform at the output. The amount of distortion will obviously depend on the frequency ω at which the input pulse is incident. If ω is low enough, none of the harmonics are much attenuated and so there is no distortion. This is indeed the case up to about $f = 200$ Hz ($\omega = 1256$ radian s^{-1}). Looking at the corresponding waveform (Fig. 4.21b), it may be seen that there is already a slight smoothing of the Gibbs phenomenon. At $f = 800$ Hz (Fig. 4.21c) the smoothing is quite considerable, and at $f = 3.2$ kHz (Fig. 4.21d) the higher harmonics have been entirely eliminated.

The physics is quite different but the conclusions are essentially the same as those we drew in Section 4.2, where we discussed the result of frequency-dependent attenuation upon the displacement of a vibrating string. It is true for transmission lines as well as for vibrating strings that if higher harmonics have higher losses then, eventually, we end up with the fundamental component only.

It is not the aim of this course to pass judgement on transmission lines and modulation techniques. It must be clear, however, from the calculations above that it is not advisable to launch pulses on our transmission line. The distortion is too big even for moderate pulse repetition frequencies. The obvious modulation technique is, of course,

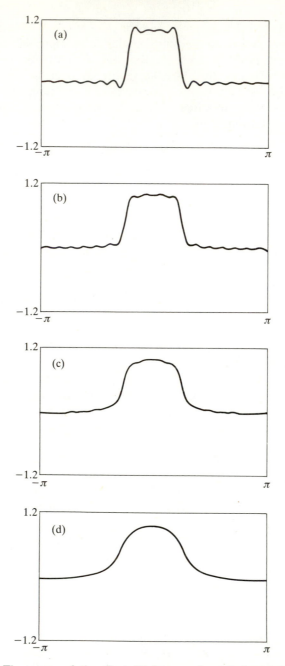

Fig. 4.21 (a) The sum of the first 17 harmonics of the periodic voltage pulses launched at the input of the transmission line, $x = 0$. (b), (c), and (d) The sum of the first 17 harmonics of the output voltage arriving at the end of the transmission line for $f = 200$ Hz, 800 Hz, and 3.2 kHz, respectively.

frequency division multiplexing, but that is really the subject of another course.

Exercises

4.1. A uniform string is fixed at $x = 0$ and $x = l$. The initial velocity is zero and the initial displacement is of the form

$$y(x, 0) = \delta x(l - x), \qquad (4.124)$$

where δ is small. Plot $y(x, t)$ as a function of x for discrete values of t covering a whole period.

4.2. A uniform string is fixed at $x = 0$ and at $x = l$. The initial velocity is zero and the initial displacement is of the form

$$y(x, 0) = \delta x^2(l - x). \qquad (4.125)$$

Find the temporal variation of the displacement at the midpoint, $x = l/2$.

4.3. A uniform string is fixed at $x = 0$ and at $x = l$. The initial displacement is zero and the initial velocity is given in the form

$$y_t(x, 0) = c_0 \sin \frac{\pi x}{l}, \qquad (4.126)$$

where c_0 is a constant. Plot $y(x, t)$ as a function of x for discrete values of t covering one period.

4.4. A uniform string is fixed at $x = 0$ and at $x = l$. The initial displacement is given in the form

$$y(x, 0) = \delta x^2(l - x), \qquad (4.127)$$

where δ is small. The initial velocity is also given as

$$y_t(x, 0) = c_0 \sin \frac{\pi x}{l}, \qquad (4.128)$$

where c_0 is a constant. Plot $y(x, t)$ as a function of x for discrete values of t covering one period.

4.5. A uniform string is stretched between two points a distance l apart. One end ($x = 0$) is fixed and the other end ($x = l$) is moved in a direction perpendicular to the string. Its displacement as a function of time, $y(l, t)$ is given by the triangular function shown in Fig. 4.22, where δ is small. Find the forced response.

4.6. The ends of a rod (at $x = 0$ and at $x = l$) are kept at zero

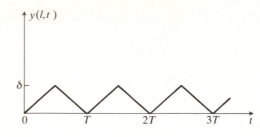

Fig. 4.22 Displacement at the end of the string as a function of time

temperature. The initial temperature distribution is equal to

$$u(x, 0) = u_0\left(1 - \cos\frac{2\pi x}{l}\right), \tag{4.129}$$

where u_0 is a constant. Find $u(x, t)$.

4.7. The ends of a rod (at $x = 0$ and at $x = l$) are kept at zero temperature. The initial temperature distribution is equal to

$$u(x, 0) = u_0 \sin\frac{\pi x}{l}, \tag{4.130}$$

where u_0 is a constant. Find $u(x, t)$.

4.8. The cooling of a sphere of unit radius is governed by the differential equation

$$u_{rr} + \frac{2}{r}u_r = u_t, \tag{4.131}$$

where r is the radial coordinate. The initial condition is

$$u = 1 \quad \text{at} \quad t = 0 \quad \text{for} \quad 0 < r < 1 \tag{4.132}$$

and the boundary condition is

$$u_r + u = 0 \quad \text{at} \quad r = 1 \quad \text{for} \quad t > 0. \tag{4.133}$$

Find a solution of the differential equation of the form

$$u = \frac{1}{r}R(r)T(t). \tag{4.134}$$

Show that the temperature distribution is given by

$$u = \frac{8}{\pi r}\sum_{k=0}^{\infty}\frac{(-1)^k \sin\left[(2k+1)\pi r/2\right]}{(2k+1)^2}\exp\left[-\frac{(2k+1)^2\pi^2 t}{4}\right]. \tag{4.135}$$

4.9. The attenuation for a particular transmission line of 70 km length is given by eqn (4.122), where $\omega = 2\pi/T$ and T is the period. Assuming that the input voltage at $x = 0$ is of the form of a triangular function of amplitude V_s, plot the output voltage for $N = 40$ (the first 40 Fourier components) for $f = 1/T = 400$ Hz, 1.6 kHz, and 6.4 kHz.

4.10. In a well-insulated telegraph cable (a very low-frequency transmission line) we may assume that $L = 0$ and $G = 0$. The end $x = 0$ of such a cable is kept at constant voltage V while the end $x = l$ is kept earthed for a time sufficiently long to ensure that the voltage at every point of the cable has assumed the steady-state distribution $V(x)$ corresponding to these conditions. Suddenly, at time $t = 0$, the end $x = 0$ is connected permanently to earth. By obtaining a Fourier series for the function $V(x)$, find the voltage distribution along the cable at any subsequent time t.

Plot the voltage as a function of space for judiciously chosen discrete values of time. Can you find any similarity with the curves shown in Figs. 4.18a and b for heat conduction?

Appendix

The Gibbs phenomenon

THE AIM here is to have a more critical look at the Gibbs phenomenon illustrated by one particular example. We shall choose for our function an odd square wave function defined as

$$f(x) = \begin{matrix} \pi/2 \\ -\pi/2 \end{matrix} \quad \text{if} \quad \begin{matrix} 0 < x < \pi \\ -\pi < x < 0. \end{matrix} \qquad (A.1)$$

The corresponding Fourier series may easily be found in the form (already derived in eqn (2.42) but needing to be multiplied by $\pi/2$)

$$f(x) = 2 \sum_{i=1}^{\infty} \frac{\sin(2i-1)x}{2i-1}. \qquad (A.2)$$

We shall now plot $f_N(x)$, the sum of the first N harmonics for $N = 17$ and 29 in Figs. A.1a and A.1b respectively in the interval 0 to $\pi/2$. As expected, the Gibbs phenomenon is there all right—the series approximation overshoots the original function. For even higher values of N (=41, 53, 65) the corresponding series are plotted in Fig. A.1c in the interval 0 to $\pi/10$. The tendency is clear: as N increases, the size of the first maximum appears to remain unchanged but it moves closer to $x = 0$. Essentially the same conclusions were drawn in relation to Fig. 2.9 when the Gibbs phenomenon was discussed for the function $(x/2\pi)^2$ defined in the interval 0,1. This time, however, we want to go further. Further in what sense? We have already said that, as N increases, the size of the first maximum remains the same but it moves closer and closer to $x = 0$. Is there anything more to be said? Well, let us pretend for the purpose of this Appendix that we are mathematicians. We would say then that just because there is a certain tendency shown up to $N = 65$ that does not mean that the same tendency will prevail as $N \rightarrow \infty$. We need a proof.

The first step is to find the position of the first maximum. The finite series for $N = 2n + 1$ is given by

$$f_N(x) = 2 \sum_{i=1}^{n} \frac{\sin(2i-1)x}{2i-1} \qquad (A.3)$$

and its derivative by

$$\frac{df_N}{dx} = 2 \sum_{i=1}^{n} \cos(2i-1)x. \qquad (A.4)$$

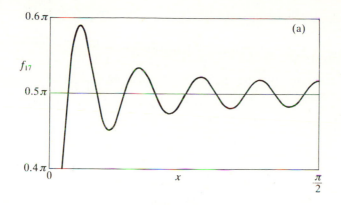

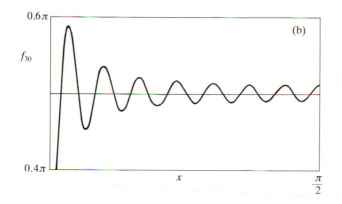

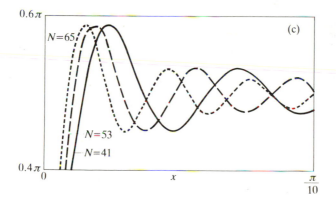

Fig. A.1 The sum of the first *N* harmonics for the Fourier series defined by eqn (A.1): (a) *N* = 17; (b) *N* = 30; (c) (———) *N* = 41, (– – –) *N* = 53, (· · · ·) *N* = 65

Luckily, this is a series that can be summed up by recognizing that

$$\sum_{i=1}^{n} \cos(2i-1)x = \text{Re} \sum_{i=1}^{n} e^{j(2i-1)x} \tag{A.5}$$

and that the expression on the right-hand side of eqn (A.5) is a geometrical series. We obtain

$$\frac{df_N}{dx} = 2 \sum_{i=1}^{n} \cos(2i-1)x = \frac{\sin 2nx}{\sin x}. \tag{A.6}$$

Note that there is no zero at $x = 0$. The limiting value as $x \to 0$ is $2n$. The first zero occurs at $x = \pi/2n$. The second derivative at this point is negative and thus we have managed to find the position of the first maximum.

Next, we need some estimate of the magnitude of this maximum. In order to find that we have to adopt the approach mathematicians use, i.e. we shall write down a number of relationships which, apparently, have nothing to do with the question under investigation but later turn out to be relevant.

Let us first integrate eqn (A.6). We obtain

$$f_N(x) = \int_0^x \frac{\sin 2nu}{\sin u} \, du \tag{A.7}$$

True to the spirit of the mathematicians' approach we shall now introduce a function

$$g(x) = \int_0^x \frac{\sin 2nu}{u} \, du = \int_0^{2nx} \frac{\sin v}{v} \, dv \tag{A.8}$$

and express the difference $f(x) - g(x)$ in the form

$$f(x) - g(x) = \int_0^x \sin 2nu \left(\frac{1}{\sin u} - \frac{1}{u} \right) du$$

$$= \int_0^x \sin 2nu \frac{u}{\sin u} \frac{u - \sin u}{u^2} \, du$$

$$= \int_0^x \sin 2nu \frac{u}{\sin u} \left(\frac{u}{3!} - \frac{u^3}{5!} + \frac{u^5}{7!} + \cdots \right) du. \tag{A.9}$$

Next, we want to put a bound upon this integral so that it is smaller than a certain value. If x is within the range 0 to $\pi/2$ then we can say that within that range

$$\sin 2nu \leqslant 1, \qquad \frac{u}{\sin u} \leqslant \frac{\pi}{2} \tag{A.10}$$

and

$$\frac{u}{3!} - \frac{u^3}{5!} + \frac{u^5}{7!} + \cdots \leqslant \frac{u}{3!} \tag{A.11}$$

and the inequality

$$\int_0^x \sin 2nu \frac{u}{\sin u} \left(\frac{u}{3!} - \frac{u^3}{5!} + \frac{u^5}{7!} + \cdots \right) du \leqslant \int_0^x \frac{\pi}{2} \frac{u}{6} \, du \tag{A.12}$$

applies, leading to the relationship

$$f_N(x) - g(x) \leqslant \frac{\pi}{24} x^2. \tag{A.13}$$

So we have managed to prove that the difference between $f(x)$ and $g(x)$ tends to zero as x tends to zero.

Let us see now what happens at $x = \pi/2n$, the position of the first maximum. At that point the difference between the two functions is of the form

$$f_N(x) - g(x) = f_N\left(\frac{\pi}{2n}\right) - \int_0^\pi \frac{\sin v}{v} dv, \tag{A.14}$$

where we have used eqn (A.8).

Next, turn our attention to the integral expression in eqn (A.14). It may be shown by some more sophisticated means (with the aid of so-called path integration in the complex plane) that

$$\int_0^\infty \frac{\sin v}{v} dv = \frac{\pi}{2}. \tag{A.15}$$

Accepting eqn (A.15), it follows that

$$\int_0^\pi \frac{\sin v}{v} dv = \frac{\pi}{2} - \int_\pi^\infty \frac{\sin v}{v} dv. \tag{A.16}$$

Substituting eqn (A.16) into (A.14) we find that

$$f_N\left(\frac{\pi}{2n}\right) - g\left(\frac{\pi}{2n}\right) = f_N\left(\frac{\pi}{2n}\right) - \frac{\pi}{2} + \int_\pi^\infty \frac{\sin v}{v} dv. \tag{A.17}$$

Now comes the crunch. When $n \to \infty$ the value of x tends to zero. If x tends to zero then, according to eqn (A.12), the difference between $f_N(x)$ and $g(x)$ tends to zero. Hence we may make eqn (A.17) equal to zero, which will yield

$$\lim_{n \to \infty} \left[f_N\left(\frac{\pi}{2n}\right) - \frac{\pi}{2} \right] = - \int_\pi^\infty \frac{\sin v}{v} dv. \tag{A.18}$$

But this is just the amount of overshoot, the amount by which f_N exceeds $\pi/2$. All that is left to do is to integrate the right-hand side of eqn (A.18). This can only be done numerically. I believe it was first done about a hundred years ago. The value is

$$\int_\pi^\infty \frac{\sin v}{v} dv = -0.2811.$$

Thus, at its first maximum the Fourier series takes the value of $\pi/2 + 0.2811 = 0.589\pi$ at $x = \pi/2n$ in the limit of $n \to \infty$. It is the value we can see in Fig. A.1.

Index

boundary conditions 83, 99

chopped sinusoidal 13
circuit theory 68, 71
circuits
   RC 77
   RL 68
   RLC 69, 70
coefficients
   arbitrary period 31
   derivation of 15–19
   exponential 50
coordinate transformation 25
current 67, 77, 104

differential equations
   first-order 63–4
   higher order 65–6
   non-linear 74–6, 77
   ordinary 62–78
   partial 79–113
   second-order 64–5
differentiation 29
discontinuity 19, 28

electrical engineering 12, 49, 53, 63, 67–74
even functions 17
exponential form 49
exponential function 26

filters 54–9, 109
Fourier integral 45
full-range 38, 60
function
   exponential 26
   sawtooth 24
   square wave 22, 23, 34, 35, 68
   triangular 23, 97, 111

Gibbs phenomenon 11, 14, 31, 37, 40, 58, 69, 114–17

half-range 38, 60, 86, 89, 95
Hamming window 55, 57, 61
heat conduction 97–104

initial
   conditions 83
   displacement 95, 111
   velocity 95, 111

Kirchhoff's law 105

laser 101
least-squares method 4
losses 88, 90–5

mechanical engineering 63

non-linear differential equations 74–6, 77

odd functions 18
Ohm's law 67, 104
optimization 38
optimum approximation 37–8
ordinary differential equations 67–78
overshoot 11

Parseval's theorem 53, 60, 67
partial differential equations 79–113
propagation coefficient 106
pulses
   rectangular 8, 51

quarter-range 38, 45

RC circuit 77
rectangular pulses 8, 51